T0122778

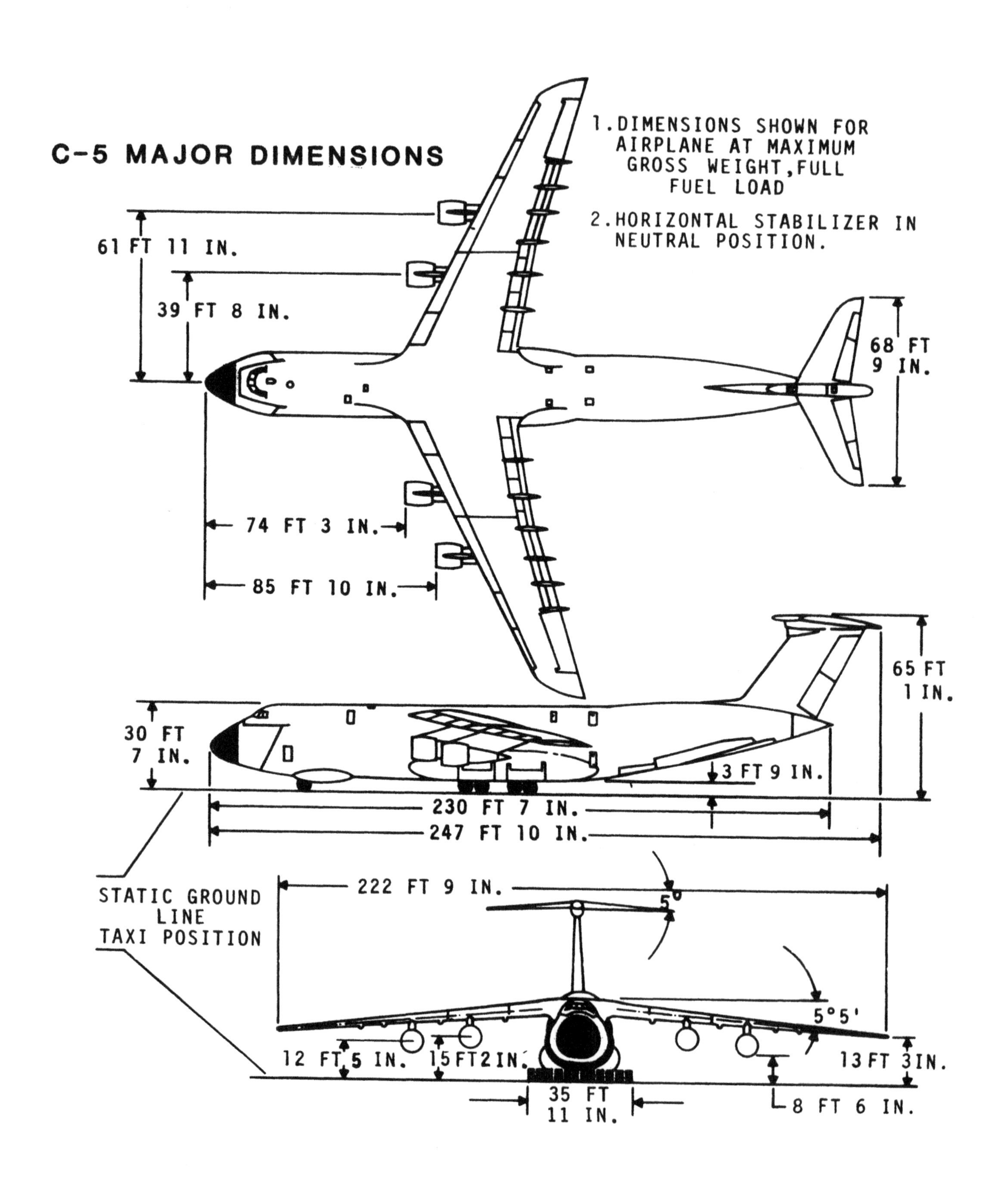
C-5 MAJOR DIMENSIONS
1. DIMENSIONS SHOWN FOR AIRPLANE AT MAXIMUM GROSS WEIGHT, FULL FUEL LOAD
2. HORIZONTAL STABILIZER IN NEUTRAL POSITION.
61 FT 11 IN.
39 FT 8 IN.
68 FT 9 IN.
74 FT 3 IN.
85 FT 10 IN.
65 FT 1 IN.
30 FT 7 IN.
3 FT 9 IN.
230 FT 7 IN.
247 FT 10 IN.
222 FT 9 IN.
5°
5°5'
STATIC GROUND LINE TAXI POSITION
12 FT 5 IN.
15 FT 2 IN.
13 FT 3 IN.
35 FT 11 IN.
8 FT 6 IN.

BASIC DATA	WING	VERTICAL	HORIZONTAL
AREA, SQ FT	6200 *	96 .067	965.824
ASPECT RATIO	7.75 *	1.24	4.74
TAPER RATIO	0.371 *	0.800	0.370
SWEEP AT 25% CHORD	25° ●	34°56′	24°35′
ROOT CHORD, IN.	545.303	371.197	250.160
CHORD PLANFORM BREAK	336.366	———	———
TIP CHORD, IN.	84.036	296.958	92.559
MAC IN. (PROJECTED)	370.523 ◆	335.452	183.438
AERODYNAMIC SPAN, IN.	2630.437	414.256	811.620
ANHEDRAL	5°5′32″ **	———	5°0′
INCIDENCE	3°30′ ***	———	+4° TO -12°

◆ (TRUE MAC = 371.209) (REF)

● SWEEP OF OUTER WING IN ROOT CHORD PLANE

* IN AERODYNAMIC PLANE

** IN AIRPLANE FRONT VIEW, 3° 30′ 0″ AROUND ROOT CHORD

*** ROOT CHORD

WEIGHT ●	
CONDITION	LB
EMPTY	318,469
DESIGN FLIGHT GROSS (2.5g)	728,000
USEFUL LOAD - DESIGN FLIGHT	409,531
MAXIMUM DESIGN GROSS	769,000
USEFUL LOAD - MAXIMUM DESIGN	450,531

● GUARANTEE FIGURES

(4) TF39 ENGINES

MISCELLANEOUS DATA		AILERON	SPOILERS	TRAILING EDGE FLAPS	LEADING EDGE SLATS	ELEVATOR	RUDDER
AREA SQ FT	INBOARD	———	218.128	493.407	◐ 322.348	180.159	UPPER-99.
	MIDDLE	———	72.600 △	161.576 □	⊖ 99.147	———	———
	OUTBOARD	252.792	140.004 ▲	336.799	◓ 227.044	78.515	LOWER-127.
	TOTAL/SIDE	126.396	215.366 ⊘	495.891	324.269	129.337	TOTAL-226
DEFLECTION		UP 25°	GRD 60°	TAKE-OFF 25°	SEALED 21.5°	UP 25°	± 35°
		DOWN 15°	FLT 22.5°	LANDING 40°	SLOTTED 22°	DOWN 15°	

△ ACROSS WING BREAK (LATERAL CONTROL)

▲ OUTBOARD WING (LATERAL CONTROL)

⊘ ALL SPOILERS (GROUND OPERATION)

◐ INBOARD WING (SEALED)

⊖ WING BREAK TO OUTBOARD ENG

◓ OUTBOARD ENGINE TO WING TI

□ ACROSS WING BREAK

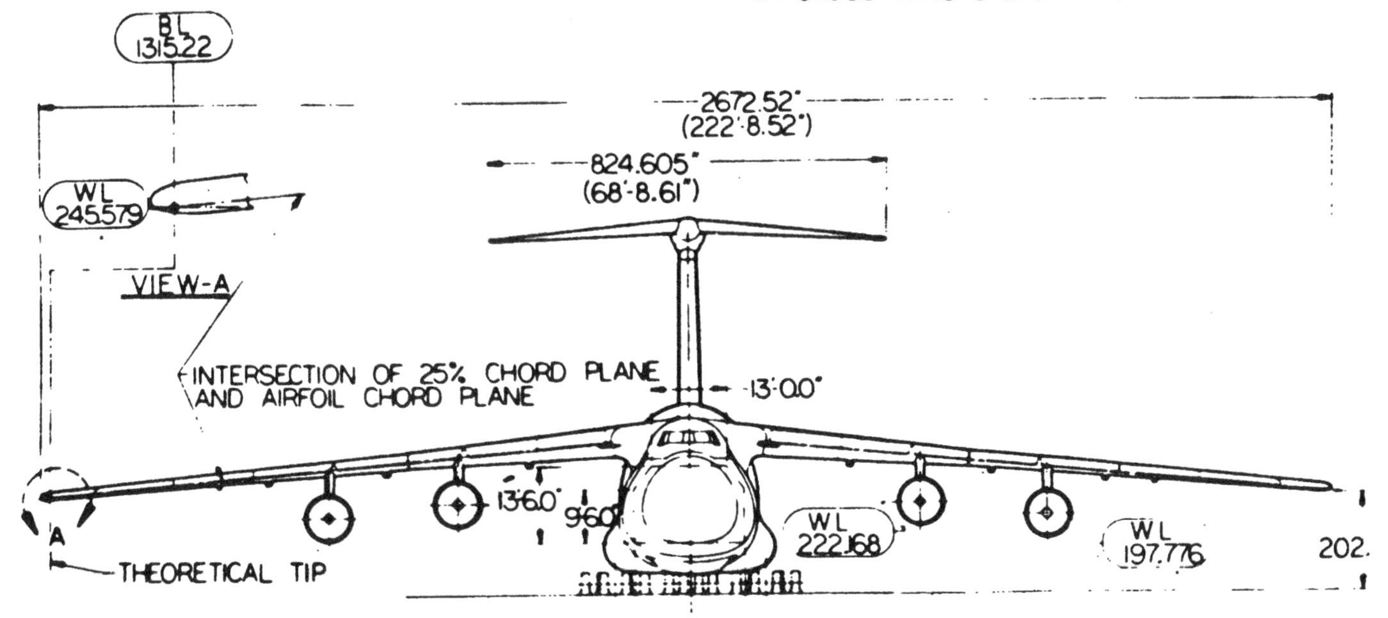

The C-5 Galaxy History

Crushing Setbacks, Decisive Achievements

By Roger Launius & B.J. Dvorscak

Turner Publishing Company

Dedication

This book is dedicated to pilots Bud Traynor, Til Harp and their crew of sixteen.

Shortly after departing Saigon in April 1975, with 340 Vietnamese refugees and 50 Americans aboard their mercy mission C-5A, a catastrophic failure resulted in the loss of normal flight control at 25,000 feet. Despite the apparent hopeless situation, this crew remained calm and professional, guiding the badly crippled giant to a semi-controlled crash landing in marshland. Miraculously, more than half of those aboard survived. The heroism of the crew was immediately repeated. After escaping from the wreckage of the cockpit, they rushed to assist survivors in the detached fuselage.

Turner Publishing Company

Turner Publishing Company Staff:
Dayna Spear Williams, Editor

Library of Congress Control Number: 2001090215
ISBN: 9781681624389

Additional copies may be purchased directly from the publisher. Limited Edition.

Contents

C-5A flies over Stone Mountain, Georgia, July 1968.

Acknowledgments

This text on the history of the C-5 airplane is the product of the collective efforts of numerous individuals over several years. Dr. Roger D. Launius, NASA Chief Historian, did the initial research detailing why the military cargo airplanes came to be a factor in fundamental national military posture. His work is the backbone of this text. A decided shift in form will be noted by the serious reader as the material presented comes, as it did, from a flight crewman's point of view.

Other individuals are also due recognition for their extensive effort and continuous encouragement. Indeed, it may read like a roster from GELAC of the 1960 through the 1980. Lockheed Georgia pilots Frank Hadden, Glenn Gray, Leo Sullivan, Ric Johnstone, Walt Hensleigh and Jesse Allen. USAF pilots Joe Schiele, Dave Wilson, Joe Gutherie, Ev Howe, Til Harp, Rich Dorre, and Larry Phillips. Flight engineers George Garger, Charlie Callahan, Jerry Edwards and Jimmy Campbell. Flight Test Engineers Mitt Mitendorff, Bill Harris and Alan Youngs.

Tom Disney, Air Loads Department manager in the Structures Division, realigned the text describing the structural ground and flight testing elements for qualifying the C-5A. Engineers Paul Horton, Jack Paterson, Charlie Posey, Ellis McBride, Bob Meyer, John Bennett, Mike McCarty, George Gelly, John Edwards, Sam Darden, Ed Rossman, Jim Saylor, Al Shewmaker and Bill Grosser all responded to interviews and telephone requests with valuable information.

Bill Nichols, author and University of Georgia professor, reviewed an early draft.

More than a handful of very senior management assisted in my search for historical insight. Bob Ormsby, Lloyd Frisbee, E.B. (Gibby) Gibson, Eddie Gustafson, Bard Allison, Bill Perreault, Charlie Ray, Lee Rogers, Ev Hayes, Whit Holland, Ed Shockley, Bob Christopher, and Bill Arndt all patiently contributed to this effort.

Four individuals are due special recognition:

- Frank Wilson, GELAC's Fluid Dynamics Division Manager during the early C-5A contract years, who described the innovative design effort required to address the aerodynamic challenges defining contract performance guarantees.
- Vern Peterson, a fellow pilot, who reviewed the entire text (resulting in spirited pilot commentary).
- Phil Sullivan, Lockheed Corporate VP, he completed a total, concentrated review to enhance structure of the text.
- Scott Barland, Senior Aerodynamics Specialist, who reviewed technical and writing details; and, provided assistance at a moments notice. A special thanks.

PREFACE

The Lockheed C-5 Galaxy is truly an effective military aircraft. This airplane was designed to airlift outsize military cargo into areas of armed conflict. It has size and strength to carry the heaviest equipment in the US Army's inventory, including the main battle tanks, such as the M-1 Abrams or the M-48, which gross 120,000 and 108,000 pounds each. Since the Galaxy can deliver these tanks in pairs, in "ready for immediate combat configuration", this singular capability gives the Galaxy a unique trademark.

The C-5 was first flown 30 June 1968 from Dobbins AFB, in Marietta, GA; and placed in service at Altus AFB, Oklahoma in December 1969. The true potential of the airplane was demonstrated early in its operational life. In 1971 the C-5 delivered new CH-47 helicopters to and recovered damaged helicopters from Cam Rahn Bay, Vietnam. Additional airlift missions moving outsized cargo to strategic locations world-wide followed. The C-5's special airlift capability was emphasized in an overwhelming manner during "Operation Nickelgrass". This crucial military airlift of war material to Tel Aviv in October 1973, enabled the United States to successfully re-supply the Israeli armed forces in their fight for survival against the Arab States.

In this thirty two day airlift operation from the United States to Tel Aviv, Israel, the C-5A aircraft airlifted 10,600 tons of equipment, 48% of the total tonnage delivered by all aircraft. The C-5 required only one third of the total fuel consumed by both types of aircraft involved, and completed its flights in only 25% of the total mission hours flown by the Military Airlift Command airplanes required for this operation.

The following year, in August 1974, The C-5A air-launched the Minuteman missile down the Vandenburg test range. The Galaxy is the only airplane in the world capable of this task at this time. Many aviation records have been set by the C-5 in its thirty-two year history. In addition to its military airlift prowess, the airplane has been extremely effective in bringing medical and survival equipment to earthquake, tornado, and hurricane victims around the world.

In the early years of operation, the C-5 program faced severe problems, several of which were life threatening to the success of the C-5; but, they were managed successfully by a determined company and USAF leadership. Born under the now infamous total package procurement contract (TPPC), design and development were accomplished during the politically charged environment of the Vietnam war. Despite all the obstacles, eighty one of the C-5A airplanes were produced and placed into service. A second contract was signed in 1984 for the production of an additional fifty C-5B airplanes, the last of which was delivered in 1988.

The Galaxy remains an essential and successful component of this country's military airlift inventory.

C-5 Flights – Significant Operations

30 June 1968	First flight of C-5A, Dobbins AFB Marietta Ga.
17 December 1969	First production C-5A delivered to Altus AFB.
December 1970	C-5 airlifts a B-57 wing to Thailand. The C-5 was the only aircraft capable of carrying a wing of this size. Airlift reduces B-57 down time by five months.
January 1971	C-5 delivers three CH-47 Chinook helicopters to Cam Rahn Bay,Vietnam. The helicopters are airborne 10 hours after arrival. C-5 returns three damaged CH-47s to New Cumberland Army Depot in Harrisburg, PA.
8 to 13 June 1971	C-5 demonstrates short field performance at Paris Air Show. Average gross weight 500,000 pounds; landing distance to a full stop, 1500 feet.
5 April 1972	C-5As airlift one million, six hundred thousand pounds of war material, including 98,000 pound M-48 tanks to Vietnam in ten missions.
14 October 1973 - 15 November 1973	C-5s airlift 10,673 tons of war material to Tel Aviv, Israel, re-supplying war material losses during the Arab-Israeli conflict during Operation Nickelgrass.
8 October 1974	C-5A successfully air launches an 86,000 pound Minuteman missile down the Vandenberg AFB range. This is the heaviest single package ever airdropped
22-31 May 1976	Typhoon disaster relief flights to Guam. 24 C-5A and 83 C-141A and 3 C-130 aircraft airlift 2,650 tons of medicine and supplies.
19 June 1977	C-5A delivers a 40 ton super-conducting electro-magnet plus forty tons of support equipment to Sheremetyevo Airport in Moscow. The 5,124 mile non-stop flight began from Chicago, O'Hare airport. (Photo on page 139.)
27 Nov. 1978	C-5 airlifts computerized axial tomography scanner to Algeria for immediate medical use by President Houari Boumedienne. C-5 is the only airplane capable of delivering equipment in "ready to use" configuration.
29 September 1981	C-5 airlifts eight F-5 aircraft to Prince Hassan Air Base, Jordan on a non-stop flight from Dover AFB
24 September 1983	C-5 s/n 0060, severely damaged on landing at Shemya AB,Alaska was flown gear down, non-stop to Dobbins AFB, Marietta, Ga. Three in-flight refuelings were necessary. Lands 25 September after 15.8 hrs.
1984	C-5s flew 459 Special Assignment Air Missions (SAAM).
17 December 1984	C-5 sets world records. Largest payload, 245,371 pounds, air-lifted to 2,000 meters. On same flight, using IFR, recorded highest weight any aircraft has ever flown to date—920,836 pounds.

The C-5A has airlifted complex heavyweights:

- A 152,000 pound A-7D flight simulator.
- A Titan IIIC missile from Denver to Patrick AFB.
- Two M-60 battle tanks weighing a total of 196,000 lbs.
- A SKYLAB laboratory.
- A 62,000 pound ATLAS/Centaur launch vehicle.
- A 100-bed mobile Army hospital from Ft. Hood to Nicaragua.
- A US Navy 57 ton main propulsion reduction gear from California to Maine.
- 136,200 pounds of cargo, non-stop, from McClellen AFB, California, to Iran; 210,385 statute mile flight, two aerial refuelings, a 20 hour and 40 minute flight.

The C-5 Galaxy is the only aircraft capable of airlifting the new Army SM-XM-1 tank, which measures nearly 12 feet wide, 32 feet long, and weighing 59 tons. The C-5 can airlift two tanks anywhere in the world. (MAC Publication, 1984.)

10 September 1985	First Flight of C-5B.
22 September 1986	Philippine Airlift authorized by Congress. Two C-5s airlift 174,185 pounds of medical supplies worth 10 million dollars to Manila.
1988	C-5 modified for space shuttle support flies after modification.
7 July 1988	Boeing contracts with DOD for C-5 airlift of a CH-47D helicopter and 64 passengers to Tibet China for Mount Everest Centennial Expedition.
11 August 1988	C-5 establishes record for heaviest airdrop. Four Sheridan tanks and 73 combat equipped troops airdropped in two passes. Total weight airdropped was 177,000 pounds.
9 June 1989	C-5 airdrops 190,346 pounds during OPERATION RODEO at Pope AFB, N.C. This breaks record of 177,000 pounds set by a C-5 on 11 August 1988. Four Sheridan recon vehicles were dropped on one pass followed by 73 paratroops on the second pass.
5 October 1989	C-5 delivers a TACAN navigation station to MacMurdo AB in the Antarctica.
7-9 August 1990	Desert Shield/Desert Storm, Phase I, deploying and sustaining forces to defend Saudi Arabia.
11 November 1990- 10 January 1991	Desert Shield/Desert Storm, Phase II, deploying offensive forces capable of evicting Iraquis from Kuwait.
11 January 1991- 28 February 1991	Desert Shield/Desert Storm, Phase III, support wartime operations.
1 March 1991- 15 August 1991	Desert Shield/Desert Storm, Phase IV, redeploy forces, sustain in-country forces, humanitarian operations.

C-5A 0001 in taxi tests prior to its first flight.

Chapter 1

In its early years, the military airlift system's capability was a poor second to civil flight organizations of the day. The C-5A reversed that adverse comparison.

Origins of the C-5A Concept

The concept of the C-5A "Galaxy" of the mid-1960s was conceived more than a decade earlier by leaders of the United States Air Force (USAF). Like other USAF weapons systems, the C-5A was a product of the evolving aviation technologies in the 1950s. As military airlift activities wound down following the Korean conflict, the Military Air Transport Service (MATS) found itself embroiled in a debate with segments of the commercial aviation industry and members of Congress over the role of the military airlift in both peace and war.

To many, much of the traffic over the MATS strategic airlift system of routes appeared more appropriately belonging to the private sector, especially when MATS' pilots flew essentially the same routes as commercial airlines. Intense competition among the scheduled and supplemental commercial carriers (1) in the uncertain airline market of that era had created a situation by the mid-1950s that appeared threatening to even the most financially sound airline. The CEOs of the various airlines saw a lucrative market in the Department of Defense (DOD) for cargo they could carry, and therefore wanted a much larger slice of MATS' airlift business. Moreover, there was great public interest in reducing the expenditures and size of the federal government, and a move from a military to a commercial contract airlift system for much of DOD's cargo could potentially yield significant total savings.

In this environment, Congress showed sustained interest in the relationships between military and civilian air transport operations. The first formal congressional discussions were in the 1956 House Defense Subcommittee hearings. Disturbed by the Army's inability to deploy its strategic stateside forces to foreign theaters, as well as by questions raised by the Hoover Commission on military air transport activities and their possible infringement on civil carriers, Representative Daniel Flood (D-PA) oversaw a series of airlift hearings. During the presentations, Representative Flood criticized MATS' use of outmoded, propeller-driven aircraft and recommended the acquisition of new jet aircraft to accommodate the Army's air transport requirements. In his view, large modern aircraft, designed solely for military use, were needed. The airplanes would be capable of transporting the Army's troops and heavy equipment together, thereby ensuring the timely arrival of cohesive fighting forces. (2)

Plans for modernizing the MATS fleet did not come to fruition until after the continued attention of Congress forced senior DOD officials to consider the problem anew during the latter 1950s. Even so, little of substance actually took place until the 1960-1961 time period when three critical actions occurred. The first involved a study conducted at the direction of President Eisenhower. He asked DOD Secretary Neil McElroy to examine the role of MATS in all environments. Completed in February 1960, "THE ROLE OF MILITARY TRANSPORT SERVICE IN PEACE AND WAR" contained the first national policy statement on military airlift. The report had nine provisions. It directed that commercial carriers, through the Civil Reserve Fleet program, would augment the military's need for airlift. MATS, in turn, would provide "hard-core" military airlift. The provisions further stipulated that MATS would undergo modernization to fulfill its military requirements and proposed joint civil-military development of a long-range, turbine-powered cargo aircraft (3). These were critical elements leading to the requirement for new, large cargo aircraft.

The second action arose when Representative Carl Vinson (D-GA), chairman of the House Armed Services committee, asked Representative L. Mendell Rivers (D-SC) to head a special subcommittee to look into the Army's requirements for airlift in support of the increasingly important flexible response strategy. (4) As early as 1951, the Army's leaders had been harping on the need for strategic airlift deployment capability and had asked the USAF to be capable of airlifting a tactical airborne assault force of two and two-thirds divisions, plus one additional division, to potential combat theaters worldwide. Cargo weight requirements per division were placed at 5,000 tons for movement to established facilities, and 11,000 tons for austere locations. Just to deploy 5,000 tons earmarked for one of these divisions was estimated as requiring 272 C-133-type aircraft, the planes then commonly used by MATS. (5) During the Rivers hearings, the Army Chief of Staff, General L.L. Lemnitzer, restated the Army's request as requiring sufficient airlift to move the combat element of a division within 14 days and two divisions within four weeks. It quickly became apparent to Representative Rivers that USAF could neither support these requirements, nor did it have any realistic plans underway to reach those goals. The result was a stinging rebuke to both the Joint Chiefs of Staff and the Department of the Air Force for failing to create enough capability in MATS to meet potential contingencies.

The third critical action occurred during the presidential election campaign of 1960 when Senator John F. Kennedy (D-MA) made the airlift issue a part of his political campaign. His goal of flexible response for the Nation's defense strategy required the ability to project military power throughout the world. He even spoke of the need for developing "additional air transport mobility; and obtaining it now" in his State of the Union address in January 1961. Thus, rapid mobility became a key element of the Kennedy Administration's posture for deterring conflict (6). This presidential position on airlift was carried to fruition when the rapidly rising defense strategy of flexible response, both in nuclear and conventional arena, gained preeminence among the nation's leaders. An able advocate of flexible response, Kennedy's Secretary of Defense, Robert McNamara, pressed forcefully for a jet transport's acquisition, sometimes in the face of opposition, and always experiencing a submerged sense of apathy from Headquarters USAF. (7)

Believing the near term need called for a medium sized, work-horse transport the DOD under McNamara emphasized the consummation of a program that had been first started in a very small way in 1959. Working closely with the Army, the C-141 "Starlifter" was designed to carry about 70% of an airborne division's equipment a distance of 5,500 nautical miles at airspeed of 500 miles per hour. Built by the Lockheed-Georgia Company, the C-141 revolutionized air transport for the American military both in terms of speed and capability. (8) It was not simply a military version of a commercial airliner; past acquisition efforts in that direction had always possessed serious drawbacks, such as loading and unloading armored equipment. It did generally utilize proven technologies. However, its configuration embodied a series of genuine "firsts" in the aircraft world, many of which were later embodied to good effect in the C-5A.

For example, all cargo aircraft with capability to airdrop cargo, prior to the C-141, incurred a severe fuselage drag penalty due to the boxy shape of the fuselage after-body, which included the aft cargo doors. This boxy design reduced range, cruise speed, and payload. The uniquely streamline-shaped after-body of the C-141 fuselage caused no drag penalty at all, when compared to that of conventional passenger transports with no cargo airdrop capability. Indeed, the overall configuration concept of a transonic, swept wing, T-tail transport with fan-jet nacelles under the wing was a brand new art form in the American skies. *When copies of this configuration shape later appeared both in Russian and a Japanese design, the C-141's place in aviation history was assured.*

It should be added, however, that the support for added airlift had come largely from outside the DOD. While certain Army leaders were advocating more airlift, they perceived it largely as a means of deploying paratroopers and still regarded surface transportation as their primary mobility system. Likewise, the USAF was not firmly committed to additional airlift, with the general exception of officers in MATS or airlifters who had moved to other positions in the USAF. The reasons for this lack of concern were complex. Though airlift was officially considered one of the primary missions of the service, most Air Force officers did not accept it as coequal with missions performed by fighter and bomber aircraft. Airlift, in essence, did not really fit into the scheme for optimal use of air power. It remained a stepchild, an auxiliary force, not contributing directly to the quest for air superiority or bombardment. Although it was important, perhaps the impression that it was closely tied to an essentially unglamorous logistical effort reinforced the stepchild position of airlift. In addition, the perception that airlift was tied to the Army probably determined the importance it was assigned in USAF Headquarters circles. The divorce from the Army in 1947 had been a difficult one, and the Air Force had sought to show how it had a mission and significance beyond that of supporting ground operations. (10) For air transport acquisitions to be successful, therefore, sufficient Congressional and key executive branch interest had to be developed to counteract the pervading apathy of most USAF leaders.

Even as the development program was underway with Lockheed for the C-141, USAF leaders, prompted by senior civilian officials, began to work on the acquisition of another jet transport, much larger than the C-141. Well aware of the problems with earlier transports and the need to develop a huge cargo aircraft to accommodate both combat troops and bulky outsize (11) equipment, on 9 October 1961, MATS issued a "Qualitative Operational Requirement" for the replacement of the only other outsize airlifter in the Air Force, the C-133 "Cargomaster"(12). This preliminary effort was followed by Air Force requirements document in June 1962. The C-5 was initially designed to carry payloads of 200,000 pounds, twice the capacity of MATS current largest aircraft, and it was specifically required to have a much wider and taller cargo compartment. The Army also wanted the new airlifter to be capable of airdrop operations and austere field landings and takeoffs. With such a large aircraft, the Army would no longer have to leave behind a sizable portion of its firepower when moving by air. Aerial refueling gave the larger aircraft unlimited range, enabling the projection of forces to any part of the globe. (13)

Changes in the size of the Army's equipment during the next two years gave further drive to MATS request for a large jet transport, which could move outsize cargo. Accordingly, in March 1964 the DOD approved a Specific Operational Requirement (SOR) document for the CX-HLS Experimental Cargo Aircraft. (The tag HLS means "heavy logistics system"). (14)

The engineers of the Air Force Systems Command's Aeronautical Systems Division (ASD) at Wright Patterson Air Force Base, Ohio, had firm ideas on the size and performance of the aircraft. They envisioned that it would have four turbofan jet engines, pylon mounted on swept wings, would fly at high subsonic speeds; would be capable of aerial refueling; could airdrop men and equipment and be able to carry 100,000 pounds of cargo. Also, the airplane was to have the capability of being loaded from the front and rear of the fuselage, a feature popularly called "Drive on, Drive off." They also viewed the airplane as 210 to 240 feet long, with a wingspan of 215 to 233 feet. The cargo area would be 17.5 to 19.5 feet wide, 13.5 feet tall and at least 120 feet long. These dimensions made the CX-HLS the largest aircraft in the world. They were estimating the gross weight to be between 650,000 and 725,000 pounds. A unique feature for the new transport was the use of 12 to 16 low-pressure tires on each main landing gear and four to six on the nose gear. Such high flotation gear would enable the aircraft to land on most "unimproved" airfields. Initial takeoff performance at home base would require a field length of less than 8,000 feet. Operations at unimproved sites of 4,000 feet were also required. This would bring "transports closer to the battle areas" and would "eliminate the middle man handling both in deployment and re-supply." With its ability to use support area airfields, the USAF felt that the new aircraft would in effect quadruple the number of usable, existing airfields in the free and contested areas of the world. (15)

Preliminary studies for development of this aircraft had been submitted by several aircraft companies in 1963, and in June 1964 Headquarters USAF selected three potential prime contractors for the new transport: Boeing, Lockheed-Georgia and Douglas Aircraft. All three were among the top ten aerospace firms in terms of assets and contracts in existence. Furthermore, each had extensive experience with jet transport development and construction. Thus, DOD had good reason to believe that each would be able to successfully complete the contract. Boeing had brought out the 707 in the 1950s and it made jet passenger operations commonplace. Not long thereafter Douglas had entered its own design, the DC-8, into the jet passenger plane sweepstakes in the early 1960s, and captured a portion of the market. Lockheed, which had considerable experience as a military transport builder, at that time was developing the C-141 for the USAF and, presumably, this new, jet cargo plane could be a larger version of that craft. (16) Indeed, Lockheed's senior management viewed the C-5 competition as an opportunity to expand on its C-141 experience, to improve MATS' capability to airlift outsize cargo and carry all Army equipment required in an emergency. (17)

These three firms received study contracts for preliminary conceptual work in the summer of 1964. They delivered initial designs in September, and by year-end the Air Force was prepared to move into the contract definition phase of the project. On 22 December 1964, after conferring with President L.B. Johnson, Defense Secretary McNamara announced on national television the decision to develop the C-5, then the world's largest aircraft. His statements included the plan for initial buy at 50 of these giant transports, with a future purchase to acquire more than 100 transports. Combining these with the C-141 fleet would increase MATS' airlift capacity by more than 600 percent by 1970. He quoted development costs over a five-year period as about $750 million, of which some $150 million would be included in the fiscal year 1966 budget. In addition, although he said nothing about it, another $1.5 billion was anticipated for procurement of the aircraft after development. (18)

With the decisions made, USAF Headquarters issued a request for proposal (RFP), including a 1,500-page document of detailed specifications for the aircraft. Each company submitted proposals and received contracts for the preparation of the final bids. In total, these companies had more than 6,000 people working on C-5 proposals in early 1965, preparing design, engineering, production, and cost specifications for the Air Force. Discretely competing teams of designers and builders of airframes, engines, and all types of support equipment services were assembled. Because of the complexity of this contract and the depth of study it required, the USAF subsidized the preparation of bids, providing a total of $61 million for these studies. (19)

Competition was keen. The size of the contract, the money it generated, and the advantage construction of a new wide-body military transport would give the firm in its competition for a civilian airliner, were great prizes. Fortune magazine remarked that Boeing, Lockheed and Douglas were all "aware that the stakes were appreciably greater than the program itself. The winners could expect to get a corner on the commercial market for a plane that promises eventually to become a standard workhorse of the air transport business."(20) Because of this benefit, as well as other intangible forces, the competition was said to be one of the most strenuous in aerospace history.

The competition was all the more stiff because Lockheed, the eventual builder of the C-5, was in a struggle for survival and had to pull out all the stops to win the contract. While the other firms were serious about their proposals and very much wanted to receive the contract; their situations were not as critical. For example, both Boeing and Douglas were relatively well balanced in both commercial and government contracts and did not have to rely on any one sector for their survival. Lockheed, on the other hand, was relying on military sales for about 90 percent of its business. This was partly the result of remarkable success it enjoyed producing military transports. Its C-130 "Hercules" developed in the early 1950s was exceptionally effective as a tactical airlifter. More than 1,000 were procured by DOD, and hundreds more were sold to other nations. But the extremely rugged characteristics that made it totally appealing to military requirements posed a weight penalty in the commercial aviation fields; thus only a few dozen were sold commercially. (21) Lockheed also designed and built 284 C-141 Starlifters, which are excellent airplanes but are not well suited to commercial use. None were used by the commercial sector. The production run for the C-141 was also nearing its end during the C-5 competition. Without a C-5 follow-on contract Lockheed would have been forced to lay off thousands of workers. Success in the C-5 competition was thus critical to Lockheed's well being. (22)

Lockheed's new Chairman of the Board, Daniel J. Haughton, was committed to obtaining the contract for his company. During this time period the government procurement system was rapidly changing; a situation which Haughton followed closely. The changes were extensive, beginning in 1961, after R.S. McNamara became Defense Secretary. While he was not committed to reducing expenditures in weapons procurement for his department, he was committed to obtaining better value for the money expended. "More bang for the buck" became an informal watchword in the Pentagon during McNamara's administration. He had come to government from Ford Motor Company, bringing with him skills and values of the automotive industry. He intended to place the DOD on an economic basis and judge its results. McNamara brought in an army of cost analysts and program experts to manage weapons procurement. His "modus operandi" was to negotiate a tight contract with very little "wiggle room" for the contractor and to cut down on cost over-runs which always received bad publicity. If Detroit auto manufacturers could build vehicles within a more or less well defined budget, which included design, engineering, research and development and production, then he planned for defense contractors to do the same.

Congressional Influence

Lockheed understood well the emerging role of Congress as an oversight agency for big-ticket government contracts. Until the 1960s, Congress had essentially been involved in defense policy only to the extent that certain exceptionally influential members wanted to be involved. Those few individuals established most of the funding priorities and marshaled forces to support or defeat certain programs as they deemed appropriate. Obtaining the key supporters in Congress was the most effective means of ensuring an easy passage of legislation affecting the program. Not to win that support ensured the program's demise. While this continued to be the case for the era in which the C-5 was procured, this barony system was beginning to fracture and consensus became harder to obtain. While Representative Mendell Rivers (D-SC) had the clout to organize support in much the time-honored manner, the old order was beginning to wither away as a younger generation of Congressmen were less willing to accept the counsel of their senior members. (25) An example of such change was the actions of the then junior senator from Wisconsin, William Proxmire, who opposed the C-5 and, on a couple of occasions, came close to having Congress zero out the funding.

Proxmire told a Joint Economic Committee hearing on defense policy in 1969 about his apprehensions in acquiring the C-5. "If we get enough C-5s," he commented, "and if they should happen to fly once they are made, we could send large numbers of troops anywhere overnight. If we have big planes, which will, on a moment's notice, take two or three divisions to every outbreak that may occur, wherever it may be, we will be tempted to do it. But I don't think we should be projecting our military power all over the world, and trying to settle every quarrel that breaks out anywhere. I do not think we have the wisdom and the experience or the manpower to run the world, and to keep the peace in that sense"(26). Despite the shortsightedness of this remark, the fears of which were not borne out in subsequent years, Proxmire was representative of a new breed of Congressmen less willing to accept the leadership of senior members. He willingly spoke out with all manner of minority positions.

The first three months of 1965 were critical in the contracting process devoted to "Project Definition" with all the three potential contractors. On 20 April 1965, Lockheed, Boeing, and Douglas each brought in their technical proposals. Their cost estimates were delivered the following week. This ended the Project Definition phase and started the review and decision phase. The Air Force established a review group of 450 analysts from numerous organizations to review and score the proposals. The winner was to be announced in the fall of 1965.(27)

The C-5 program, as it developed in the early 1960s, was an important step forward for the USAF, as well as providing Lockheed unique economic and technological opportunities. The purchase of the C-5 fit the new national policy of flexible response strategy perfectly, and thus, became a linchpin of its success. Without adequate airlift to move troops to trouble spots around the globe, any conventional capability was a hollow force. The civilian leaders of our nation and the Army understood this very well. By the early 1960s, the USAF establishment accepted this philosophy also. Naturally MATS leaders were delighted with this new attitude as it generated both procurement of the C-5 and a strengthened status for military airlift.(28)

The technology of the C-5 was also to be evolutionary rather than revolutionary, but the result was a radical – maybe revolutionary – change in the manner in which airlift was regarded and utilized by the American military.

The realization of the ability to airlift large cargos over intercontinental distances into either combat or non-threatening positions in a matter of hours was the revolution the C-5 fostered. It led to an entirely new avenue for the employment of airlift, its maintenance and logistical support systems.(32) While the acquisition process was agonizingly slow, the technology evolutionary, and the politics at times desperate, the result was a fundamental leap forward in airlift capability and responsiveness. That reality had to wait for the execution of the C-5 contract.

Lockheed Georgia flight line.
Lockheed photo.

Chinook loading onto a C-5A. Lockheed photo.

Chapter 2

Planners will, if allowed, improve a concept ad infinitium.

Military Requirements – Lockheed's Response

Part I: Establishing C-5 Design Requirements

With the issuance in October 1961 of MATS' Qualitative Operational Requirement (QOR) statement for a replacement to its aging fleet of C-133 turboprop heavy-lift cargo aircraft, the C-5 emerged as a legitimate concept. The QOR, as described in chapter 1, was refined and made more explicit in the USAF's Specific Operational Requirement(SOR), outlined in table 2-1. This document, issued in 1963, started Lockheed's pre-concept formulation process.(1)

This SOR was the result of a great many exhaustive USAF studies covering a range of operational considerations. The driving force was that U.S. military planners had recognized that rapid response strike forces were simply not functional until they had most of their fighting equipment with them wherever they went. Previously, planners had relied on pre-positioned military supplies, located strategically prior to any combat, near potential trouble spots around the world. As the decade of the 1960s began to unfold, the pre-positioning concept began to look more and more hopeless: there were too many trouble spots. Also, even with pre-positioned equipment in storage 50-100 miles away from where it was needed, the time involved to get it, make it operative, then take it into action, was thought to be prohibitive for too many scenarios.

Accordingly, the concept was accepted that the airlift of troops to anywhere in the world required that combat equipment must go with them, arriving functionally ready for action. This included not only personal equipment, but trucks, tanks, helicopters, and special equipment such as bridge launchers. Many loading and unloading studies, which considered the nature of all classes of army equipment, began to define the required cargo compartment dimensions as well as the airplane's payload. These dimensions would allow the C-5 to efficiently transport ALL U.S. Army vehicular equipment, both present and planned. The airplane envisioned by military planners was evolving into a very efficient transporter of conventional cargo that also had outstanding capabilities for airlifting of very large, heavy military equipment into demanding austere environments.

To improve their mobility and cohesiveness, the U.S. Army prefers to pack as much divisional bulk cargo and equipment as practical in their vehicles.

Table 2-1: Major SOR Requirements

Payload:	100,000 to 130,000 pounds
Cruise Performance:	440 Knots at 30,000 feet
Runways:	Take-off in 8,000 feet at maximum gross weight Landing in 4,000 feet with 100,000 pounds cargo
Personnel Capability:	Galley and Latrine for 25 people for 15 hours. Oxygen for 25 people for 5 hours
Cargo Compartment:	Length – 100 to 110 feet Width – 16 to 17.5 feet Height – 13.5 feet
Cargo Pallets:	Two rows palletized cargo 88 inches, or one 88 inch and 108-inch row.
Cargo Loading:	Straight through, door on both ends.
Floor load Height:	Truck bed level
Power Plant:	Six Turbofan engines
Reliability:	90 percent probability of completing 10 hour mission
Availability:	No later than June 1970

Typically, three-fourths of a mechanized division's bulk will be transported in this manner. Therefore, only about one-fourth of the division's cargo actually travels on pallets. Thus, much of the bulk cargo can be transported on outsize and oversize loads as "topping," making use of available payload capability and space which is insufficient for vehicle loading. Therefore, very few bulk-only sorties are required, which improves airlift efficiency. The U.S. Army had also found that the probability of executing a successful operation is greatly enhanced by maintaining unit integrity. This makes it necessary that vehicle crews accompany their vehicles to facilitate loading, unloading, and regrouping operations. This meant that the airplane, in addition to its load of equipment, must accommodate not only flight and vehicle crews, but have space and payload available for combat troops. It is obviously desirable that this added provision be met without encroaching upon the usable cargo volume. Figures in chapter six illustrate how these critical design features were incorporated into the C-5A.

Many years of USAF and industry experience had highlighted the need for self-sufficiency in military airlifters. The capability to airlift to forward airfields close to mission objective areas, then return to staging locations, all without reliance on ground support equipment, could prove essential to the success of the overall mission.

Items to be considered in providing this self-sufficiency included:

- All aircraft ground handling and maneuvering must be accomplished without the aid of tugs or other external equipment.
- Routine maintenance, such as tire changes, must be completed using on-board tools.
- Vehicles must be loaded and unloaded by driving on and off at ground level without the use of depot provided platforms.
- Power from ground based systems will not be available for starting aircraft engines or operating other aircraft equipment.

These specifications would lead to the integration of hitherto unprecedented design features, such as the "knight's visor" front cargo door, and a kneeling capability, which lowered the floor of the parked aircraft to truck-bed height.

Airdrop missions are almost always performed only when conventional landing and unloading is nearly impossible or because of close proximity of the enemy. During airdrop, the risk to the airplane itself is great. Of course, no battlefield commander would risk one of his heavy lift aircraft in a hazardous airdrop mission if he has an alternative. Further, the jumping of a large number of paratroopers and airdropping vehicles from a single aircraft is certain to cause more dispersion of the delivered personnel and material than having the same items delivered by a fleet of smaller aircraft in formation. Generally, then, airdrop missions would likely be accomplished by the smaller, less expensive aircraft such as the C-130 Hercules. Aerial delivery capability was, nevertheless, required for the C-5, particularly for extremely heavy vehicles, such as full sized battle tanks, which could only be airdropped from a very large airplane. The requirement for palletized aft loading and unloading was emphasized since, obviously, parachute extraction can only be made from the rear of an airplane in flight. Also, the airflow environment around the airplane after-body with open cargo doors and paratrooper exit doors must not be turbulent or in a way that generates hazards to the men or material exiting the airplane. Lastly, the airplane response to the center-of-gravity transient movement occurring during the extraction of the heavy cargo items must be easily controllable by the pilot and not present any flying qualities hazards.

Selection of Cruise Airspeed

Virtually every design decision on an airplane is a compromise between differing basic considerations. For example, structural weight, aerodynamic drag, and procurement dollars versus a desired performance, which includes payload, landing and takeoff distance, flying qualities, and cruise speed. Nowhere is this better illustrated than in the selection of the required cruise airspeed. Trade studies were run to assess the total system cost of a fleet of heavy lift aircraft. One such cruise speed trade study in which Lockheed participated is described as follows.

The defined task to be accomplished by the C-5 was deploying, within thirty days, one Reorganized Infantry Division from Ft. Lewis, Washington, and one Reorganized Mechanized Division from Ft. Campbell, Kentucky, to Bangkok Thailand.

Costs then were:

A) Procurement of sufficient aircraft to complete the plan
B) Operation of these aircraft five hours daily for ten years on an equivalent design mission during peacetime.

A mixed force of C-141 and C-5 airplanes was assumed where the heavy, outsized cargo would be airlifted by the C-5.

Parametric airplanes were "designed" for cruise Mach numbers of 0.75, 0.767, 0.80, and 0.85. Each of these used a 17.5 x 13.5 x 100 foot cargo compartment, a 4000 nautical mile mission, a design payload of 130,000 pounds, and a maximum structural capability of 150,000 pounds for the payload. Airplane characteristics and relative costs are summarized in Table 2.

Table 2

Airplane Version	"A"	"B"	"C"	"D"
Mach Number	0.75	0.767	0.80	0.85
VARIABLES				
Wing Area, Sq. Ft.	4,775	4,680	4,600	4,550.
Wing Sweep Angle, degrees, at 25% Chord	20	25	30	37.5
Gross Weight, pounds	598,000	610,000	628,000	658,000
Empty Weight, pounds	240,500	243,000	249,900	259,900
Fuel + Crew, oil, baggage	227,500	237,000	248,600	268,100
Std Day T.O. Thrust Pounds	153,000	171,600	184,200	216,000
Number of airplanes	104	100	98	94
Ten year system cost	1.0	1.01	1.02	1.07

This study showed that an effect of increased cruise speed is an increase of aircraft empty weight (EWE). Hence procurement costs increase, as do fuel requirements and operational costs. There turned out to be many other more subtle penalties to be paid for a cruise Mach number higher than Mach 0.767.

One such relates to balance desired between directional and lateral control power design requirements. The single, most critical directional control requirement, for virtually all multi-engine aircraft, is the need for continued flight, climbing on a straight course, after one engine fails upon take-off. This high level of rudder control power needed for controlling asymmetric thrust after take-off with a single engine power loss, however, becomes a severe liability in the event of a control system failure known as a rudder "hard-over," where the rudder moves suddenly and uncontrollably to full deflection and remains in that position. This causes a sideslip motion of the airplane, which, in turn, produces a large and possibly uncontrollable rolling moment. The greater the wing sweep angle, the greater the rolling moment thus produced, and incidentally, the less effective the ailerons become. Many swept wing transports in service today will roll uncontrollably in the event of a rudder hard over under certain cruise flight conditions, no matter what action the pilot takes. However, NOT ONE of Lockheed transport airplanes suffers from this risk.

Aircraft propulsion system engineers are fond of reciting to their airframe system counterparts, "no airplane progress takes place until new and better engines are provided for them to use", in other words, "without us you are helpless." In the case of the C-5, they were absolutely correct. Referring back to the above discussed parametric (trade) study, three turbofan engine families were considered. They are briefly compared in table 3.

Table 3

Engine Choice	Pressure Ratio measure of power	Takeoff Thrust Sea Level Std Day, pounds
Pratt/Whitney JT3D	1.26—1.37	21,000 to 23,000
Pratt/Whitney STF200C-4	1.95	21,500 to 39,500
General Electric GE6	3.0	20,000 to 40,000

The relative system cost for designs using six JT3D-8A and six STF200C-4 engines were compared. The JT3D-8A is the engine used in the C-141. The GE6 engine family was almost identical in performance to the STF-200C-4 family, so that aircraft design and cost characteristics were the same for either engine. The new engines had higher thrust, higher pressure ratios, higher bypass ratios, and better thrust to weight ratios than the earlier JT3D, so the minimum cost for using these engines was approximately 5 percent less than the minimum provided by the JT3D, in spite of the additional development cost.

It was no surprise that the new, much more powerful, high bypass ratio (BPR) engines showed a significant reduction in system costs. Only four were required per aircraft versus six of the smaller engines. The higher BPR meant lower fuel consumption, which, in turn, allows a lighter, less expensive airframe as well as a reduction in operational fuel costs. General Duane H. Cassidy, Commander of the Military Airlift Command in the mid-1980's recalled the process.

"We started to build the C-5 and wanted to build the biggest thing we could build. We went to the engine manufacturers knowing that the power plants were going to determine how big the airplane could be. We said, 'How big an engine can you build? ' Herman the German, a guy working for GE at the time, Gerhard Neuman was his name, there was a book on him. He was a bald headed guy, very heavy accent, first generation German. He said, 'We can design you an 8:1 bypass ratio engine.' Everybody rolled their eyes because that was twice as much as anybody had ever done before, but he designed it and built it. It was the TF-39 engine. That engine then has become a core for GE derivatives that are flying in most every aircraft in the world today."

Structural Considerations

Even if one were content to ignore improvements in the engineering state-of-the-art in going from one airplane program to the next, the idea of "scaling up" a smaller airplane to a larger configuration simply doesn't work. The reasons lie in some seldom discussed scaling laws. Before we begin discussion of how Lockheed responded to this rapidly developing requirement for this giant airplane, it is worth considering two extremely profound, physical barriers which were being challenged by this concept of the C-5. The first is called the "square /cube law". In essence, the law states that stress in similar structures increases with linear dimensions if the imposed load is proportional to the structural weight since the latter grows as the cube of linear dimensions while the material cross section carrying the load increases only as the square.

For a simplified, but factual illustration of this principle, consider the familiar water tower found in many small towns. Suppose the city manager of one such town needed a larger water tower than the one the town has in place. He tells his staff to pull the blueprints of the existing tower, double all the dimensions, and build it. For our basic example, the reader is invited to view the original water tower as a cube, 10 feet on each side, positioned 25 feet above ground by a single square column, 1 foot on each side. Also assume the entire structure is made of steel plate that is 1/4 (one quarter) inch thick. The tank thus holds 1,000 cubic feet of water, approximately 7,481 gallons, which weighs 62,400 pounds. The weight of the structural steel in the tank and its supporting column weighs another 8,000 pounds which means that 24 square inches of steel in the supporting column is loaded to a stress level of 2,933 pounds per square inch(psi). For the smaller tank this is assumed to be an acceptable margin of safety.

Now, double all the dimensions. We now have a 20-ft cube tank that holds 8,000 cubic feet (just under 60,000 gallons) of water weighing just under half a million pounds. The supporting column has increased to 4 ft on a side, the steel plate is 1/2 inch thick throughout. Structural weight has grown to 56,600 lbs. The support column has doubled in cross-sectional area to 96-sq. in. but is having to support 555,783lbs or nearly 8 times the total weight of the original design. The stress level in the supporting column has doubled to 5,789 psi. Without delving into material properties and engineering design allowables, one can see how the gut reaction approach of scaling up to a larger size will not produce a safe structure without considerable advances in material strength.

Flying Qualities

The other or second fundamental law of physics, which inherently penalizes larger moving vehicles of all types may be referred to as the cubed/fifth power law. This law simply notes that the aerodynamic, or driving moments, which act on an aircraft during a maneuver vary as the cube of its scale while the inertial or internal resisting moments to any maneuver vary as the fifth power of its scale. We observe this law in the familiar fact that a small boat operating at a given speed is capable of achieving a turn in many fewer boat lengths than can a large ship operating at the same speed. Another example, mockingbirds can harass hawks with impunity during flight because their smaller size allows them to so out-maneuver the hawk; he is helpless to defend himself from their attack.

The flying quality requirement for the first Wright Flyer purchased by the U.S. Army was as follows: "This machine must be capable of being flown by a reasonably intelligent man after proper training." Somewhat more sophisticated criteria were developed from research performed at Ames Aeronautical Laboratories. Since these criteria are illustrative of the overall inertia problem related to aircraft flying qualities, we will examine a few of them in some detail.

A flight situation is postulated where an aircraft is on approach for landing, the pilot notes he is below his target glide slope by an amount that requires correction. To decrease his descent rate, he must pull back on the control column, causing the airplane's elevators to deflect upward, pushing the tail downward and rotating the nose of the airplane upward. That elevator deflection initially causes the airplane to increase its descent rate, because of the download it must impose on the tail to cause a nose up rotation. It is only after the airplane has begun rotating nose-up, in its secondary response to the elevators deflection, that the rate of descent will begin to decrease as the wing angle of attack increases. The time lag between the initial control surface deflection and airplane response in the desired direction is critical to acceptable flying qualities. The NASA-AMES criterion of interest here states that if the time lag between control input and the beginning of desired airplane response exceeds two seconds, the airplane will not have satisfactory flying qualities. Factors which influence the airplane's ability to meet this requirement include size, effectiveness, and rate of movement of the elevator, tail length, the lift curve slope of both the horizontal tail and the wing, the moment of inertia of the airplane in the pitch axis, and finally the airplane's dynamic response.

The C-5A has an unusually long tail length, expressed as 4.2 "mean aerodynamic chord" lengths. This gave the desirable effect of the longer lever and assisted in meeting the NASA-AMES requirement without having to resort to an unnecessarily large elevator and horizontal stabilizer.

Thus, for better or worse, the package of customer requirements were for what few realized was an entirely new class of aircraft. It would provide unprecedented mobility to U.S. strike forces, but at a price of entering into the unknown in ways imperfectly understood.

The Definition Process

Design of the C-5 transport occupied four sequential phases:

1. Pre-concept formulation.	October 1961 to April 1964
2. Concept formulation	May 1964 to December 1968
3. Project definition	December 1964 to October 1965
4. Acquisition	October 1965; tests complete 1972

Many parametric and other design studies were conducted during the first few years. Using configuration, size and number of engines, structural materials, control systems—every conceivable option was investigated with the objective of minimizing system acquisition costs for this new class airplane. Finally on 12 December 1964, three months after Boeing, Douglas and Lockheed submitted final reports on the CFP studies, USAF issued an RFP for the aircraft. The requirements included every feature shown to be desirable in the CFP studies. The main requirements affecting aircraft design are listed below:

Payload – Range

Payload, lbs	100,000	200,000	265,000
Range, nautical miles	5,500	2,700	2,700
Limit load factor(g)	+2.5/-1.0	+2.5/-1.0	+2.25/-0.0

TAKE-OFF over 50 ft. at Sea Level at 89.5deg F.
At Basic Design Gross Weight (2.5g)..............8,000 feet
At Max Design Gross Weight (2.25g).............10,000 feet

LANDING over 50 foot obstacle at Sea Level at 89.5 deg F
With 100,000 pounds payload and fuel to return,
at midpoint of 2,500 nautical mile radius mission...........4,000ft

INITIAL CRUISE ALTITUDE Basic design gross weight 30,00 feet

LONG RANGE CRUISE SPEED 440 knots true airspeed

PROPULSION Four turbofan engines, sea level static thrust of 40,000 pounds each

Cargo Provisions

Cargo compartment size:	min width 17.5-ft. min length 120 ft. excluding ramps.
Floor area:	min. 2300 sq. feet, excluding ramps; 2,700-sq. ft. including ramps.
Height minimum:	13.5 ft. (13 feet min. width at 13.5 foot height)
Cargo accommodations:	Compatible with 463L ground and aerial delivery systems.
Forward ramp:	Full cross section exposure. Ramp angle 11 degrees.
Aft ramp:	Straight in opening 13 ft. wide and 9.5 feet high.
Clearance normal to ramp:	12 feet., Ramp angle 13.5 degrees.
Floor height for loading:	between 48 and 54 inches.

Passenger Provisions

6 bunks, 15 passenger and 87 troop seats above the cargo.

Reliability

90 percent of aircraft dispatched must reach their destination without a major subsystem failure. An additional 8% may suffer failures, which do not cause mission abort. A reliability level of 87% based on subsystem failure is to be demonstrated during the Category II (USAF) test program.

Maintainability

Quantitative requirements based on a minimum operational availability of 75%.

Landing Gear

Flotation of 94% take-off and landings without airfield repair, on Support Area Airfields (M-8 landing mat on CBR 4 sub-grade) at gross weight for landing with 200,000 pound payload, fuel for 1,000 nautical miles range, tire deflection 40%. Cross wind pre-positioning from parallel to +/-20degrees. The airplane must be capable of 180 degree turns on a 150-foot wide runway.

Operating Weight

Operating weight included pallets, troop seats, and cargo restraint devices.

Stability and Control

MIL-F-8785 and CAR requirements must be met. In addition, a rate of roll of eight degrees (8) in one second in the approach configuration at 1.2 times the stall speed at lightweight was required.

Airframe Life

30,000 flight hours of anticipated usage, including 6% at 300 feet altitude at 350 knots IAS using terrain avoidance. 12,000 landings, 5950 pressurization cycles.

A few days of analysis indicated that these requirements were mutually incompatible. The specified 40,000-pound thrust engine was insufficient to permit meeting the payload-range requirements with the large cargo compartment, the inclusion of pallets, troop seats and cargo restraint devices in the operating weight, and other requirements. Subsequently over 500 changes to the RFP were issued between 24 January and 25 March 1965. These resulted in largely eliminating the conflicts in the RFP, but the engine thrust was still somewhat too small. Lockheed firmly resolved to solve this problem and did so, partly by a policy of applying a 5% reduction to the Operating Weight (OW) estimate. The OW was estimated by extrapolating from the weight of previous smaller airplanes. An rather optimistic analysis indicated that we could expect to advance the state of the art sufficiently in structural design to reduce this extrapolated weight target by 5%. Lockheed senior management would live to seriously regret this decision.

Part II: Lockheed's Response

As noted in the previous material the results of all the design studies leading to the definitive C-5 airplane requirement showed that a C-141 style for the general configuration was the least expensive means to meet the Air Force objectives. This fact gave Lockheed a decided advantage since the C-141 development experience was applicable to a considerable extent. Lockheed's "T-tail" low drag after-body design was a major contributor to the winning design. The twenty five degree, aft-sweep, wing allowed the use of a high wing loading while retaining good short field performance and effective ailerons to provide part of the powerful roll control requirements. The under-wing pylon mounted nacelles were also a design feature which Lockheed engineers were well versed in applying.

However, many changes from the C-141 design were essential. Much better transonic airfoils than had been used on the C-141 were available by this time. Both Boeing and Douglas had been doing considerable airfoil research, but Lockheed, due to its serious engineering manpower shortage in developing the C-141, had not kept pace in this research category. Quite candidly, as the competition began, Lockheed did not know how to design a winning aerodynamic wing shape. Fortunately, however, Lockheed's aerodynamics personnel knew someone who did. At that time period, probably the greatest transonic airfoil designer in the free world was Mr. Herbert Pearcy in the British National Physical Laboratory. So, Mr. Pearcy and one of his associates were invited to visit Lockheed for a week to teach key members of the company's aerodynamic staff how to apply his "peaky" airfoil design principles. He helped set up Lockheed's first computer aided wing design program. Lockheed learned this crucial technology so late in the program that it had no time to do any transonic wind tunnel model tests on its C-5 entry wing at all prior to its competition design submittal.

Normally, this might not have been a problem since the drag substantiation within such design proposals usually relied heavily on estimated data. The C-5 contract, however, had a new feature intended to discourage optimistic aerodynamicists. Each competitor was required to furnish to the Air Force a completed transonic wind tunnel test model with his proposed design. This model was to be suitable for testing in the NASA Langley 7x10 foot transonic tunnel. Under the direction of Air Force personnel, the test data would be analyzed to learn if the contractor's performance data was valid.

It was unheard of to expect that a wing aerodynamic design could be accomplished with no development testing at all – especially when using a new and untried airfoil section with which company engineers had no previous experience. Yet, because of Lockheed's late entry into the field of mathematical aerodynamic wing design, that was exactly what Lockheed would find necessary for this proposal. John Bennett, Lockheed Georgia's best theoretical aerodynamicist, was assigned to the task of leading this crucial analysis. He absolutely worked endlessly, and although he was pressured for early release of data, he never released those data until the results satisfied the requirement. The result was that wind tunnel model designers received their design data too late for models to be completed in time for Lockheed to do any wind tunnel testing prior to submitting its final design proposal. Thus, the first entry of Lockheed's design into any wind tunnel was during the evaluation testing done by NASA. Despite Lockheed's late entry into the "advanced transonic wing" era, its design was hailed by no less than Dr. Robert Whitcomb, who headed the NASA Langley team testing all the competitor wind tunnel models. He stated to Lockheed's aerodynamics design group, "Lockheed's aerodynamic wing design was far and away the best we saw. You guys designed it just right." An unsolicited comment like that from such a recognized expert in the field made it all worthwhile.

When the competition design study began, the take-off and landing field requirements were such as to lead the Lockheed design team to believe that they could repeat the C-141 experience. That is, avoid the use of leading edge slats, thus reducing weight and cutting costs. A few weeks into the final competition effort, the Air Force issued a requirements "clarification" which, in essence, considerably changed these requirements. BAD NEWS. Lockheed would have to include a wing leading edge flap on its entry after all, but its engineering staff had no experience in the aerodynamic development of these wing devices. Once again, technology had to be borrowed, and quickly. Lockheed dispatched a two-man team to Convair to beg or borrow whatever expertise and applicable data they had and bring it back for immediate use. Thus, it was that the aerodynamic design of the Lockheed C-5 leading edge flaps followed Convair's design practice. This was not the first time Convair had volunteered aid to Lockheed Georgia. Shortly after it was announced that Lockheed had won the C-141 design competition, a large package was delivered to the office of Lockheed's competition manager. It was from Convair and it contained a complete copy of Convair's (losing) competitive design submittal. The cover letter included a statement to the effect, "We have spent millions in the preparation of this report, but it is of no further use to us. We hope that it may be of some benefit to you, as you undertake the final design of the C-141".

As noted earlier, a very important element of Lockheed's conceptual C-5 configuration was its "T" tail and low drag after-body. During the conceptual design phase of the competition, Boeing did its level best to discredit this feature in the eyes of the customer, the USAF. The experience of the "T" tail BAC-111 in encountering a very high angle of attack stall from which the crew was unable to recover, was very fresh in the minds of the transport community. Although Boeing's "T" tailed 727 aircraft were then entering airline service, Boeing declared that they would never again build a "T" tail design airplane. NASA began a series of wind tunnel testing studies to discover how "T" tailed airplanes could completely avoid deep stall pitch control lockup. And many in the Air Force community began to express reservations about this design feature. Why buy unnecessary risk, was the common attitude. Clearly, this reservation on the part of Lockheed's customer had to be eliminated quickly, or Lockheed's entry in the competition would, at the very least, be severely downgraded, if not completely disqualified.

Fortunately, Lockheed had very well documented C-141 experience on this very point. Extensive wind tunnel tests had been completed examining effects of penetration into deep stall flight regime in every possible airplane configuration checking every form of flight controllability. Because Lockheed flight test management prudently intended to be doubly sure that the C-141 had no stall controllability problem, they ordered a very comprehensive investigation of flight in the high angle of attack, (the stall and deep stall), flight regime. More than 2000 aircraft stall maneuvers were flown and comprehensive data taken. Several times a week for a period of eight months, engineers from the Aerodynamics and Flight Test Departments, with management from department and division level, gathered to study the flight data. Meetings often lasted into night sessions while flight results were compared to the Aerodynamics Department's dynamic response predictions based on wind tunnel test data. Correlation was generally good. Eventually management, engineering, and pilots gained confidence that we could look our customer in the eye and tell him that there was no deep stall lockup problems in the C-141 airplane. However, the FAA and the customer required the addition of a device called a "stick shaker/stick pusher" despite the overwhelming evidence that it was unnecessary.

So, suddenly, this wealth of experience became very valuable. A comprehensive presentation on the subject was prepared by Lockheed's Flight Sciences Division manager. He and Lockheed's Chief Engineer delivered this material (lectures) at NASA Langley and Ames, to the Air Force Flight Dynamics Laboratory, to personnel at MATS headquarters in St. Louis, as well as squadron flight crews at MATS bases. The Lockheed message was straightforward: "We really do know what we are doing in this flight regime; we know what causes control degradation and what does not. And we promise you that the C-5 will have no stall recovery problems in the high angle of attack (stall) flight zone". The message was believed. And it was correct.

For reasons mentioned earlier relative to the squared/cubed law, the C-5 engineers faced a task of weight reduction in the airplane design of unprecedented difficulty. This problem was compounded for Lockheed by its management decision to further reduce their engineers' already optimistic weight estimates by another five percent. C-141 structural design and fabrication processes were now totally inadequate to accommodate the C-5 production task. Innovation became an absolute requirement to meet this situation.

Fortunately the structural engineers' state-of-the-art had progressed considerably since the C-141 had been designed. The combination of the advent of high speed computers, the evolution of improved insight into the mechanics of structures, and the development of advanced analytical techniques enabled the structures analysts to pinpoint with much greater precision exactly how much metal was required in each structural component. Airplane models for wind tunnel use were made with many hundreds of tiny holes in the model surfaces where flow pressures were measured as the model was manipulated through every required flight condition. These provided air-loads data of excellent detail, further enhancing the precision of structural analysis. Higher strength materials and fastening techniques had become available. Al Cleveland, Lockheed's chief scientist, in his 33rd Wright Brothers lecture, noted; "The DC-3 used a stressed skin and stringer type of construction similar to that used today, except that the skins were lapped and fastened with low strength, protruding-head rivets and bolts. This approach, if applied, would result in a C-5A fastener system weight of 38,500 pounds (or about 18,000 pounds using C-130 technology). The actual weight of the C-5 fastener system is only 8,500 pounds."

No item was too small to escape the rigorous weight reduction discipline. Chemical milling was introduced and used extensively so that metal structure in a variety of sizes could be reduced to the exacting thickness and contour at all points. Tapered, interference fit titanium fasteners, in lieu of rivets, were used mainly in the wings. Titanium has a higher strength-to-weight ratio than either aluminum or steel, hence a weight savings is made. The application of the interference fit fastener, a manufacturing process, enhances fatigue life. With these fasteners a higher stress level could be accepted in many locations of the wing than had been allowed in previous design practice when aluminum alloys were used. Again, this meant structural thickness could be reduced and weight reduced.

Over 20,000 pounds of bonded aluminum honeycomb structure was used. 3150 pounds of fiberglass and 560 pounds of magnesium materials are used in lieu of heavier metallic structure. A savings of 1100 pounds was made when beryllium wheel brake discs were used in place of steel discs, despite steel being much less costly. Any engineer who takes a close look at the C-5 structure will be impressed by the extensive tailoring of all the structure. Every element in the assembly is sculptured, tapered, scalloped, contoured or otherwise configured to reduce weight to the absolute minimum. These innovative practices substantially increased fabrication costs. However, other structural engineering innovations added to the overall weight control problem rather than helping it. The dynamic response of very large aircraft to in-flight gusts, combined with the higher design stress levels in structural metals, produced a potentially more damaging array of repeated stresses, thus increasing the metal fatigue hazard. Further, since the investment cost in very large aircraft had grown so huge, customer demand for improved safety and reliability grew rapidly. Thus the requirement for damage tolerant, fail-safe design came in vogue and was applied to the C-5. One way this requirement was met involved installing thin titanium straps midway between the fuselage rings such that a skin crack advancing from a failed adjacent ring will not propagate further, and the airplane can continue to operate safely until the repairs are made. Unfortunately, these straps added weight.

Design Innovations

No discussion of the C-5's uniqueness would be complete without attention to some of its design innovations. The "knight's visor" nose design, wherein the twenty feet of fuselage forward of the cockpit swings up to allow unimpeded drive-through loading and unloading, is one such design first. The capability to lower the cargo floor to truck bed height, referred to as "kneeling", significantly enhances cargo handling.

The prominent flap track fairings on the C-5 wing are a design "first," in that their "favorable" aerodynamic interference actually produces a significant cruise drag reduction over the otherwise clean wing drag value. Normally it is expected that such a fairing would penalize by adding to the total drag number. However, during the C-5 wing development this concept of "favorable" aerodynamic interference was put to good use and later patented by the company.

One of the requirements the C-5 had to meet was that it must be capable of continuous operation from soft fields. "Continuous operation" was defined as a finite number of takeoffs and landings on the soft field before ground personnel would do any surface repairs to the landing area. In the process of design of the landing gear, literally hundreds of wheel positions were studied to define how to provide the required floatation at minimum weight and cost. The final 28-wheel landing gear configuration evolved after months of study. Many believed the requirement to be ridiculous. What battlefield commander would risk an investment such as the C-5 so close to the battlefield where no paved runways would be available? The requirement was met, however.

The fewest possible maintenance man-hours per flight hour was a contractual line item. Meeting this requirement for an airplane of the complexity of the C-5 using the C-141 era technology, would have been impossible. One of the innovations on the C-5 to assist in achieving this goal is the Malfunction Analysis Detection and Recording System (MADAR). This computer-operated system alerts and assists the flight crew in responding to system malfunctions on the airplane during flight. The MADAR is a substantial reference document equipped with a video type screen for displaying text and diagrammed material. A companion printer allows the flight engineer to extract any of the extensive check list data, and system complexities are more easily managed. Trouble shooting procedures and scheduled maintenance procedures are included in the computer memory for access on demand.

C-5 cargo compartment, view from the rear; forward visor closed, crew door open (left side). Lockheed photo.

Chapter 3

Whatever Happened to "KISS"

Controversy: Contractural, Financial, Structural

The Total Package Procurement Contract

One of the most significant aspects of the history of the C-5 Galaxy was its procurement method. This involved the unique contractual arrangements developed to purchase the aircraft. The mechanics of military buying is almost a discipline within itself, especially in relationship to major Department of Defense (DOD) weapons system procurement. In these acquisition programs three critical factors come into play and rebound off each other and all the parties in a complex set of relationships. These can be identified as questions: What is the DOD buying? Why is it buying this weapon? What is the time frame and total cost of the program? The first two are concerned with product differentiation, and the last with price. These are complex questions in any purchase, but in military aircraft acquisition they are especially difficult, in part because there is not a marketplace to set the price and other parameters of procurement for such products. It all must be negotiated from ground zero for each type of aircraft. In addition, since these aircraft represent the leading edge of technology, they are, by definition, expensive.

In all of this work the DOD must negotiate a program that produces the aircraft it wants at a price it can manage, and most importantly, one that it can persuade the President and Congress to support. At the same time, the aerospace companies must have a price that will be fair and profitable. Since most military aircraft procurements generally have been for equipment that is usually beyond the technological state-of-the-art, the costs of design, test, and production are generally unknown and very difficult to estimate accurately. The forecasting of final costs for an aircraft program at its conceptual phase is a task filled with uncertainties. Malcolm Stamper, former president of the Boeing Aircraft Corporation, remarked that "locating the break-even point is like finding a will-o-the-wisp."(1) Knowledgeable individuals have concluded that it's not until 25 or 30 aircraft have actually been manufactured that production costs become predictable to the accuracy needed. (2)

All military aerospace procurements since World War II have been plagued with this complicated situation, with mixed success. DOD could use its governmental position, as well as its unique role as nearly the only customer, to force aerospace contract prices down to the lowest levels. It has refrained from this approach, for driving aerospace companies into financial straits would only weaken the nation's ability to continue in the forefront of technology. Aerospace contractors, government policy has implicitly stated, are national resources to be nurtured for the good of the country. At the same time, those companies are an important part of the national industrial base and employ thousands of workers, directly or indirectly, and their well being is correctly considered to be important to the national economy. The government tries to avoid the economic dislocation that would result from failure of one of these giant companies. The aerospace procurement process, therefore, has aimed at providing a reasonable profit to these companies at the same time that it seeks to hold down costs, waste, and obtuse procurement practices.(3)

The results of these efforts have been mixed at best. A 1962 analysis of 12 major weapons systems procurements found that their final cost averaged 220 percent above the original cost estimates. A Brookings Institution study of DOD contracts in the 1950's shows cost increases ranged between 100 and 700 percent of the quoted prices. This Brookings examination concluded that "all large military contracts reflected an acceptance of contractor estimates which proved highly optimistic." (4)

DOD was wrestling with these problems when John F. Kennedy entered office in January 1961 and brought in a new staff of senior leaders with new ideas. He appointed as Secretary of Defense, Robert S. McNamara, the former president of the Ford Motor company, perhaps the individual having the most influence on the direction of military procurement practices in the post-World War II era. McNamara was an unusual man. He was not the typical automobile executive; his background was academic, and his methods were strict, rational, and disciplined. He looked more like a professor than a car salesman. He relied on tools of scholarship — logic, mathematics, and mountains of facts—to make decisions, rather than intuition and personal judgment. Lyndon Johnson characterized him: "Brilliant, intensely energetic, publicly tough but privately sensitive, a man of great love for his country, McNamara carried more information around in his head than the average encyclopedia."(5). Other observers said of him that he was "an analyst by nature, who, when he examines a situation, strips it down to component parts with numerical relationships between them. When he asks questions of his subordinates, he calls for answers 'with numbers in them' which means to him information wrapped in realistic terms that one can act upon."(6) The be-speckled McNamara was also rudely honest and expected others to be the same way. Kennedy said of McNamara: "He's one of the few guys around this town who, when you ask him if he has anything to say and he hasn't, says 'No'. That's rare these days, I'm telling you."(7)

Whatever else he might have wanted to do when he became Secretary of Defense, McNamara was committed to making his department more efficient. He thought it could do the same or better job with less money and he set out with a vision of reform. A huge target for his efforts was DOD's procurement practices. Fortified with studies demonstrating massive cost overruns and waste in the defense acquisition program, McNamara initiated a three part campaign aimed at reform. First, he sought interim, relatively easily done changes. Next, he consciously worked to take control of the procurement efforts from the three uniformed services and to control it from his office with strict accounting procedures. Finally, he planned to begin a different type of contracting system which would promote efficiency and be less susceptible to wasteful practices.(8)

During his first two years in DOD McNamara made several significant changes and gained control of the major procurement programs from the three services. Then he set out to accomplish his third goal, to change the method of contracting. McNamara's first big target was cost-plus contracts, which had been mainstays in DOD for many years. These contracts negotiated a base price which covered all development and manufacturing costs, and then, either a fixed fee or a percentage of cost add-on to secure a reasonable profit. McNamara believed, not without justification, this encouraged waste. Contractors had no reason to try to hold their costs down since the government covered those costs at 100 percent, and they still received money on top of that which ensured a good profit. He pressed hard for another type of contract which used a fixed price rate and then provided for additional payments in various incentive categories. He wanted the contractors to accept the possibilities of both profit and loss on any DOD contract, something that flew against popular conceptions that prime contractors would suffer greatly and might be forced into bankruptcy. McNamara, who had experience with Ford in manufacturing for an uneven market, would have none of it and pressed hard for his approach. He succeeded, but only partially. When he took office in 1961 thirty eight percent of all DOD procurement contracts were cost-plus fixed fee types, but by 1963 that type had dropped to only twelve percent.(9)

McNamara also pushed for an approach to acquisition that the Air Force had tried and then largely abandoned in the 1950s: a two-step process. In step one, contract recipients were chosen as a result of a design competition in which price was not considered a factor. During this phase the contractors would be reimbursed for their design costs, but those designs extended to a sublime level of complexity. In theory, another corporation could or would take the design proposals and build the aircraft. At the completion of this stage, the second step involved the negotiation of a contract to build the design submitted.(10)

In addition, McNamara and his supporters worked hard to eliminate what was known as "buying-in," offering of an extremely low price by the contractor in order to obtain the contract. Usually contractors had been successful in persuading the government that the military-generated design changes during production added costs to a project. In some instances, aerospace contractors had used their position as sole contractor to persuade the government that cost overruns were necessary and that it had either to sink more funding into the project or to abandon the program. That represented the classic "over the barrel" position for the government; it had to put more money into a project or to kiss whatever had already been expended goodbye. In any event, this process of "getting well" was a very real problem that McNamara wanted to end. As these efforts were germinating, McNamara recruited a key person to handle procurement for DOD, and it represented a major episode in the C-5 development program. Robert H. Charles reported for work at the Pentagon in 1963 fresh from 18 years with the McDonnell Aircraft Company, seven of them as an executive vice president. His strong aerospace industry background, coupled with his zealous commitment to the spirit of McNamara's DOD procurement reforms, made Charles an especially able individual to handle this position. He became Assistant Secretary of the Air Force for Installations and Logistics and developed a whole apparatus for measuring cost-effectiveness and determining the viability of any given program. He reasoned, along lines paralleling McNamara's, that competition between aerospace corporations for military programs was imperfect. His experience dictated that companies competed only for the research and development contract for a major program. The bids for those, which he thought should be about 20 percent of the total program cost, were often extremely low, since the winner anticipated recovering losses in the first stage by adding to the production contract. Thus, he wanted to extend competition through the production contract stage.(11)

Under Charles' direction, and with McNamara's strong encouragement, the DOD developed an entirely new type of contract, the Total Package Procurement Contract (TPPC). With this approach the Air Force awarded the contract for the whole weapon system, with potential bidders submitting quotes for research, development, test, and production of the system. It was a "womb to tomb" concept, declared Pentagon wits. The winner received a contract that specified a total cost, the delivery schedule, the specifications for hardware, and the maximum profit allowed by the government. Charles believed there were no loopholes in this contract approach which would allow industry to gouge the government. The contract was divided into segments for each major task of the program, and every task had hard target and ceiling prices as well as schedules. It also included incentives for exceptional performance. During the first segment, costs had to come in between the target and ceiling or the remainder of the contract would be canceled. Charles believed this would eliminate "buying in" by industry since there was no purpose in submitting outrageously low proposals.

Robert Charles devised this contractual arrangement, he said, "because I was convinced that we are spending more than we need to on our national defense, that defense equipment was costing far too much, due to procurement methods which imposed little discipline, particularly competitive discipline, on either industry or government, and I could do something about it."(12) The C-5 acquisition program was the first to use the total package procurement concept, and it represented much that was both good and bad about the aircraft's development.

The C-5 contract that Lockheed, Boeing, and Douglas competed for in 1964 and 1965 called for the development of a huge transport. The transport was required to carry outsize cargo—tanks, helicopters, and certain other types of vehicles—the full range of equipment to outfit an army division. On some missions the airplane would be required to airlift 50 tons of payload 2,875 miles, land in a forward combat area, and subsequently take off from this dirt runway of less than 4,000 feet in virtually any weather. Everyone knew that meeting this set of requirements would tax the technological abilities of the aerospace industry; yet, it was the type of challenge that excited the engineering elite. Daniel Haughton, the president of Lockheed admitted this in 1969: "Engineering is not an exact science and we have to invent as we go along."(13)

The C-5 contract included virtually everything that might come up in the course of a weapon procurement cycle. It specified all aspects of research, development, testing, and evaluation on five experimental aircraft; an initial production run of 53 planes, (Run A), assembled as the five experimental aircraft were being tested. Design changes were to be incorporated into production line as testing continued; an optional second production run of 57 additional C-5s (Run B); ground and flight testing of each production aircraft; crew and maintenance training for six squadrons; and finally, ground support equipment and spare parts. This comprehensive document had both incentives and penalties for delays in schedule—$12,000 per day for each of the first 16 airplanes delivered late up to $11,000,000. The contract allowed a 10 percent profit for the prime contractor which was fixed. According to Charles, " It was probably the toughest contract for a major defense system ever entered into by the Pentagon."(14)

Lockheed, Boeing, and Douglas accepted the challenge nevertheless, and bid on the C-5 work under the constraints of the Total Package Procurement contract. The three bids were submitted in April 1965 in a remarkable set of proposals. In what was probably the most intense competition in DOD contracting history, the three corporations submitted 35 tons of documentation, enough to have loaded 14 DC-3s, to substantiate their bids and illustrate the designs.

Boeing's proposal was the costliest, $2.3 billion. Even so, this was 500 million less than the company's best estimates believed were necessary to complete the work. Its analysts concluded that Boeing only had a thirty percent chance of fulfilling the contract at the $2.3 billion figure. It was a gamble in which Boeing leadership was literally betting the company. Douglas submitted a bid of $2 billion for the 115 airplanes required by the DOD. Lockheed offered the lowest price for the C-5 contract—$1.951 billion, of which $177 million was profit for Lockheed. This figure was about $250 million less than DOD cost estimators thought would be necessary. The proposal was broken down as shown in Table 1, with the majority of the cost being incurred in the research and development plus Run A section of the contract.(16)

Table 1

C-5 Contract Target and Ceiling Prices

Item	Target Price	Ceiling Price
RDT&E and Run A	$1.408 billion	$1.664 billion
Run B	$543 million	$642 million
TOTAL	$1.951 billion	$2.305 billion

In 1969 a Securities and Exchange Commission investigation reported that, like Boeing, Lockheed management had reduced its cost proposal for the C-5 by 10 percent to reach the $1.951 billion mark sent forward to the Air Force. This was not an uncommon practice, as Lockheed management used their best judgment about the competitive environment to reach their proposal price. They gambled and literally staked their company's welfare on the success of their bid. They believed at the time that the bid was low enough to win the competition, but still high enough to make a profit if everything was adequately planned. The calculated figures are shown in Table 2. The 10 percent reduction, they surmised, left a healthy profit in a sharply competitive market.(17)

Table 2

Analysis of Lockheed's Cost Proposal

Actual Cost	$1.985 billion
Target Cost	$1.768 billion
Overrun	$197 million

Price Offered to the Government

Actual Cost	$1.965 billion
Less 30 percent Overrun	($59) million
Target Profit	$177 million
TOTAL	$2.083 billion
Lockheed Profit ($177 million less $59)	$118 million

After these proposals were received the Air Force Systems Command Headquarters, at Andrews Air Force Base near Washington D.C., assembled a 400 person C-5A Evaluation Group to review specifications. For the next five months it conducted its review and struggled to determine the winner. During this time there was give and take between the bidders and the Air Force clarifying details in the proposals and revising cost estimates. Of concern in the Lockheed proposal, which was an early frontrunner in the competition because of its cost and the superb track record of the company in building military transports, was the aircraft's inability to meet some recently established requirements for glide slope and landing approach performance. In September 1965 the Evaluation panel asked Lockheed, as well as Boeing and Douglas, to incorporate these requirements into its design, and ten days later Lockheed came forward with some modifications which increased the size of the wings and flaps to improve lift and drag ratios. In this effort Lockheed held firm to its cost proposal, although the external dimensions of its aircraft had increased. (18)

When all of its questions had been answered, the Evaluation Group made its recommendation to the Air Force C-5A Source Evaluation Board comprised of four general officers who would pass a final recommendation forward to the Secretary of the Air Force. They decided that the Douglas proposal was deficient in several categories and eliminated it from competition fairly early in the process. Lockheed and Boeing both had superior designs but Lockheed's cost proposal, they believed, was too low and would lead to cost overruns. Cost overruns were something to be avoided with McNamara stalking the procurement system in search of faulty practices. They finally decided to send forward a recommendation to award the C-5 contract to Boeing.(19)

Inevitably, in matters such as this, Washington leaked like a sieve and news of this recommendation quickly went back to the Lockheed-Georgia Company, which was first demoralized, but then quickly recouped and went to work to turn the decision around. A similar reaction occurred in Seattle where Boeing mobilized to manage the decision as well. It would be naive to assume that a $2 billion defense contract would not interest the Congressional delegations in the home states of these companies; both Senator Henry Jackson from Washington and Senator Richard Russell from Georgia immediately became involved. Russell, a southern Democrat upon whom President Lyndon Johnson relied to ensure party support in the Southeast, had the president's ear, and he used this fact as appropriate to aid Lockheed's cause. Later in 1968, when President Johnson visited the Lockheed-Georgia C-5A assembly line, he addressed the employees with this comment. "I would have you good folks know there are a lot of Marietta, Georgia's scattered throughout our fifty states. All of them would like to have the pride that comes from this [C-5] production, but all of them don't have a Georgia delegation."(20)

The effort to reverse the Boeing decision was not all based on power politics, although that did play a part. The Boeing recommendation within the Air Force was NOT a clear-cut decision; a minority opinion also went forward from several quarters within the Air Force opposing the Boeing selection. The source selection report noted that the Boeing proposal did not meet the minimum aircraft width established in the RFP "which limited the utility of the aircraft", that its troop compartment was inferior to the Lockheed design, and that it was not as fuel efficient as the Lockheed plane. The Military Air Transport Service, for whom the aircraft was being built, recommended going with the Lockheed proposal rather than Boeing's because it had the lowest cost and "with one rather minor exception, met all performance requirements." The command's leadership also thought the Lockheed proposal allowed for lower life cycle costs, another important factor in the long term operation of the C-5. There was great enthusiasm for Lockheed's design for roll-on, roll-off cargo handling and payload flexibility. The Air Force Systems Command also weighed in on Lockheed's behalf, stating that its proposal produced the most attractive aircraft at the "most attractive price."

In mid-September 1965 the Air Council, made up of the twelve four star generals in the Air Force, reviewed all the options proposed relative to the C-5 program and agreed by a three-fourths majority vote to award the contract to the Lockheed-Georgia Company. In doing so the Air Council agreed with the majority of those involved in the selection process, especially with the command charged with operating the aircraft, even if it overturned the recommendation of the Evaluation Board. Implicit in this decision, additionally, was a signal that the Air Force was not going to be stampeded by McNamara and his procurement reform efforts. It was willing to take a calculated risk in favor of Lockheed, even if it might mean some cost overruns.(21) This decision went to McNamara for approval and he agreed to it without hesitation, mostly because of the low cost of the Lockheed proposal. He announced the contract award on 30 September 1965. "I have directed the Air Force to proceed with the revolutionary new transport aircraft, the C-5A," he said. "This aircraft will have a gross weight of about 700,000 pounds, 350 tons at takeoff. The 'low-cost' Lockheed Aircraft Corporation has been selected to build this airplane, and, as you already know, General Electric is to build the engines."(22)

The next day a contract for the airframe was executed with Lockheed and another for the engines was signed with General Electric. Under the Total Package Procurement concept Lockheed became responsible for the entire system's performance, including its engine once it was accepted by the Air Force for installation into the airplane. It also called for a concurrent C-5 acquisition program. In this approach the airplane was designed, and construction began and was well advanced, before the completion of the test program. The first eight production airplanes were designated principals in the test program, and they were to be in the flight test program while the production models were coming off the assembly line. The design and development test programs extended from October 1965 through the mid-October 1970, although there was an ongoing low level of activity until November 1973.

Production took place as scheduled between August 1966 and January 1973 when the last of the 115 airplanes were to have been delivered. Ground and flight systems tests, therefore, took place concurrently with production through much of the C-5 program. At the height of the effort, the 115 C-5 production run would require delivery of two airplanes per month, with lead time of 22 months for constructing a single aircraft, as shown in Table 3.(23)

Table 3

Lead Time Required to Construct the C-5A Airplane

Activity	Months Required
USAF Authorization for Production	-
Placement of Forging Orders	1
Fabrication of Forging Orders by Subcontractor	6
Machining Process	5
Assembly of Main Frame into Center Body Section	3
Three Body Sections Joined	1
Pressure test, Paint and Pre-installation	1.5
Wings joined to the Body	0.5
Final assembly and Joining Tail to Body	1.5
Flight Preparation and Production Flight Test	2.5
TOTAL	22.0

The contract allowed a maximum of 10 percent of the price for profit to the company. The contract would, however, give Lockheed an opportunity to make greater profits by delivering the planes for less than the targeted figure, or penalized the company by delivering at a higher price, thus eating into the profit. This clause, however, was offset by another aspect of the contract which required the government to pay 70 percent of costs above the target price, up to a ceiling price set at 130 percent of the target.(24)

If the cost went above 130 percent of the target, there was a process in the contract which allowed for re-pricing. Although exceptionally complex, the re-pricing clause provided an opportunity to use cost experience of the 53 aircraft in production Run A as a basis for re-negotiating the costs allowed to be charged for the 57 planes in Run B. If the costs on Run A were more than the ceiling price (130 percent of the estimates), the costs over that amount were multiplied by a factor of 1.5 (2 if the costs were over 140 percent of target). The resulting figure was then used to multiply the original figure for Run B to produce a new higher target figure. The purpose of this clause was presumably to allow a contractual out for Lockheed should it incur massive and unforeseen costs. The government, therefore, would limit Lockheed's loss or profit on the contract. The formula cut both ways. An under-run (below estimate cost factor) on Run A beyond 70% would be calculated and reduce the government's cost for Run B. The purpose of the clause was to prevent either excessive profit or catastrophic loss. This "two way street" was a beneficial safeguard for both parties, especially in view of the risks of concurrent development and production of an aircraft, which, at the time, was the largest in the entire world.

Critics stepped forward to suggest that this clause defeated the purpose of the Total Package Procurement concept and the goals of the McNamara reforms. It rewarded inefficiency and waste on the part of the contractor, the critics argued. One government acquisition expert remarked that "when I heard about that re-pricing clause, I could see right away there would be trouble."(25) Problems did develop in relationship to the contract at a later period. In 1965, both the DOD and Lockheed were delighted with the direction of the C-5A procurement. The losing bidders on the contract went to work on other projects: Boeing undertook the development of the supersonic transport and the widebodied 747, and, Douglas went to work developing its competitive DC-10 widebody. Lockheed, by winning this competition, had to forestall its own entry into the commercial widebody transport sweepstakes. By the end of the year it had worked out all contractual arrangements for the C-5A with the Air Force and was beginning to get underway in the design and test of the aircraft.(26)

Factors in the Lockheed Monetary Crisis of 1970s

The worst financial crisis for the Lockheed Corporation developed in 1969. Pressures peaked in the summer of 1971 and extended for three critical years after Congress passed the Emergency Loan Guarantee Act. The Galaxy contract was a major portion of the finance problem.

On 15 November 1969, USAF/DOD announced a reduction in the number of planes being purchased from 115 to 81. This reduced the production Run B aircraft total from 57 to 23. Thus, it drastically reduced the base value to be multiplied by the cost adjustment factor and made it impossible for the cost adjustment clause to operate as it was intended. This singular act also brought into play contract termination clauses of the TPPC which required interpretations from "Solomon." None were ever agreed upon; this impasse ramped up the cost of doing business. The added "domino" effect of the TPPC reduction in purchase affected the dozens of suppliers as well. Thus, a description of the progression of events that brought about a true financial crisis, and the manner in which a sound resolution was effected, is considered relevant to the C-5 history.

In February of 1971, the Rolls Royce Corporation of England advised Lockheed that the Rolls Royce Company was over-committed financially and would go into receivership immediately, in accord with current British law. Therefore, it legally was no longer required to abide by contract requirements to supply the engine for the L-1011 passenger tri-jet. This announcement had immediate and serious consequences for Lockheed. It put the TriStar commercial airplane program in doubt in the marketplace where timely product delivery was fundamental to success. The L-1011 program was already in its sixth year with over 150 orders on the books and initial production underway. Lockheed had invested heavily in company funded engineering design studies, market evaluation, defining resource requirements for production, and placing orders with a network of suppliers, plus working the worldwide marketing activity to obtain vital additional orders.

With Rolls Royce in receivership, not only was the commercial airplane program in great jeopardy, the four large military programs carried on by Lockheed were seriously troubled as well. This combination produced a financial quagmire. The corporation was forced to take measures that ultimately put the company survival up for vote in the U.S. Congress. This devilish predicament had many details to its structure, several of which were outside routine control of any management.

Inflation

The surge in aerospace orders for new commercial aircraft, which had occurred since 1966, coincided with military demands. Labor costs and supplier lead times increased dramatically. The backlog of aircraft orders in 1964 was $8 billion; this surged to $20 billion in 1968. Lead times for aerospace materials was 9.8 weeks in 1964; lead times doubled by 1966.(1) The Vietnam buildup added new pressures starting in 1968.

In an August 1969 report, *Fortune* magazine priced the cost escalation of the C-5A program at $1.3 billion above the contract price for 115 airplanes. Inflation alone accounted for $627 million of this figure. Thus the cost increase over the original bid is 35% of the original price. Comparative data obtained by the RAND corporation on 21 major military programs showed that costs varied from 140% to 200% of the contract price. The Lockheed built C-141 Starlifter came in at 99.6% of target price. The same report covered civil programs: the Interstate highway system price cost 260%; the Raburn House office building price cost 200% of target.

Economic

The contracts for both the C-5A transport and the AH-56 rigid rotor helicopter were priced by Lockheed in a "buyers market" using cost projections based on data from 1956 through 1964. Using the cost index(CI) data for the period 1957 through 1960 as a basis, the CI increased from 110 in 1964 to182 three years later. Development and production were, in effect, accomplished in a "sellers market".(2)

Technological

H.B. Meyers, in his *Fortune* magazine article of August 1969, described Lockheed as a company that takes risks in pushing the limits of technology. Chairman Dan Haughton was quoted, "We are in the business of taking risks – the business of developing new systems. Engineering is not an exact science, and we invent as we go along". Certainly both the C-5 Galaxy and the rigid rotor helicopter, the AH-56, pushed the engineering talent of the company. The Galaxy, despite being the world's largest airplane when it became operational, had to perform multiple tasks: cruise at high altitude at high Mach number, yet be able to airdrop heavy military cargo and paratroops at slow speed; airlift into short, dirt airstrips and return without refueling. These capabilities illustrate the Galaxy's exceptional flexibility.

By 29 March 1968, Lockheed had received initial orders for 144 TriStars, with a total valuation of $2.1 billion, from TWA, Eastern Air Lines, and Air Holdings LTD. Thus Lockheed became the first company to launch into production of wide-bodied trijets. In April, Delta Air- lines placed an order for 24 planes bringing the total first order backlog to 168 aircraft. The very next day Northeast Airlines placed an order for TriStars, adding to the backlog.

Three engines had been evaluated by the customers and Lockheed, the General Electric CF-6, the Pratt & Whitney JT-9D, and the Rolls-Royce RB-211. The Rolls engine was selected. Features considered were technical merit, initial and operational cost, and growth potential. The RB-211 received the best overall rating from both a technical and economic standpoint.

Lockheed then proceeded with financing plans to cover the commercial and current U.S. government production programs. Their existing term bank credit of $125 million was increased to $400 million after reviewing funds necessary for the various programs. This revolving line of credit with a group of 24 banks was signed on 1 May 1969. It reflected cash flow predictions at that time for the TriStar buildup and delivery, as well as the 115 airplane C-5A contract, plus the other major government programs.

During 1969 and 1970 a series of disputes arose involving four of the major military programs at Lockheed. These contractual, financial, and legal disputes involved the Air Force C-5A, the Army AH-56A Cheyenne helicopter, the propulsion system subcontract for Boeing/ USAF short range attack missile(SRAM), and ship construction for the U.S.Navy. The first three of these contracts were administered using the total package procurement procedures. Continuation of these disputed defense programs while simultaneously trying for resolution placed a severe drain on company funds. Interim financing became necessary. On March 2, 1970 Lockheed notified DOD of the need for $600 million in interim financing to carry out the disputed programs pending resolution of claims. Only the $50 million SRAM claim was settled in October1970 for $20 million. Other disputes remained open through June 1971. Meanwhile, the banks expressed concern as borrowings reached $320 million. Further advances required a pledge of collateral and stock. After a supplemental agreement was signed, an added $30 million draw was made, bringing the total borrowings to $350 million.

The DOD proposed settlements for the three remaining large contracts in December of 1970. Each one involved severe monetary penalties amounting to about $500 million overall. The government proposal for the development and production contract of the AH-56 helicopter was accepted. The remaining two disputes were submitted for resolution to the Contracts Arbitration Board. Thus, on 7 January 1971, four weeks before the Rolls Royce corporation moved into receivership, Lockheed Corporation had set a course which would effect positive strides in development and production requirements for both the military and commercial programs. The Board of Directors authorized Lockheed to "bite the bullet" and agree to the government terms at the beginning of February 1971.

The Rolls Royce board decision to place the company into receivership occurred on 4 February 1971. One week later, the receiver officially notified Lockheed of the cancellation of the RB-211 contract. As a result of these uncertainties DOD chose not to finalize the agreements that had been accepted ten days earlier. These changes in circumstances also prevented the completion of the new credit agreement with the 24 banks carrying the loans for Lockheed.

Consultations began with the British government while the airline customers and Lockheed conducted a complete re-evaluation of the available engines. The Rolls Royce engine, again, was found to be the power plant of choice. In the final analysis, engine delivery was delayed a minimum of six months with a cost increase to the customer of $180,000 per engine. The British government funded continuing development and production of the engine. This action was not gratuitous; they required the strongest possible guarantee that the market for the RB-211 would be there. They insisted (demanded) that the US government guarantee the financing of the L-1011 program. Thus, in January of 1971, Lockheed, their bankers, and DOD had agreements in hand for critical finances. Before these could be executed, the Rolls Royce bankruptcy clinker was dropped into the mix of devilish detail and unhinged everything.

This provided a field day for foes of the military in general, and Lockheed in particular, both in and out of government. By their public statements claiming company cover-up, blatant mismanagement, cost overruns, and labeling the C-5A as the world's worst airplane in any category, facts in the matter became irrelevant. The critics of the airplane, led by Wisconsin Senator Proxmire and Representative Moorhead of Pennsylvania, were joined by other publicity seekers whose reasoning in public failed the "Book of Sirac test." Those who would not be swayed by the facts were also joined by others who saw the opportunity to extract considerable gain from the situation. Realizing that the US Congress was now involved and their decision was needed quickly, principals in the dispute, which was now a national issue, orchestrated major political lobbying.

One example of measures taken by an adversary of Lockheed's position was a letter writing effort by the Chairman and CEO of General Electric, Mr. Fred Borch, to President Richard Nixon. Borch's company lost in the engine competition for the L-1011. The argument he stressed was his abhorrence of the idea that Lockheed would buy an offshore engine when engines were available in this country. He conveniently overlooked the results of the second engineering evaluation, as well as the first. In the second engine review, held immediately after the Rolls Royce announcement, it was determined that an engine change at this juncture in production would add $150 million in program costs, making the TriStar totally non-competitive. The contents of his letter became public knowledge and caused the following response from Tony LeVier, Lockheed California's most experienced and accomplished test pilot.

1 June 1971

Mr. F.J. Borch
Chairman and Chief Executive Officer
General Electric Company
570 Lexington Avenue, N.Y. N.Y.

Dear Mr. Borch:

I was considerably troubled by your letter to President Nixon and others and by the related press conference relative to the proposed legislation which would provide loan guarantees to a consortium of 24 banks in support of the Lockheed L-1011 TriStar program. It is not surprising that as a long-term Lockheed employee, I am troubled by your actions. I have always have had confidence in big business and believed that it acted in good faith; but it is at best alarming that you put into circulation on a national scale information which was both false, and misleading. Even casual inquiries would have proved them so. In the current atmosphere in which the charges of mismanagement are rampant, it seems worthwhile to ask you as head of the General Electric Company to look inward and determine how you as an individual and General Electric as a corporation could have been led into this needless trap.

My reaction is not prompted by my role as a 30-year employee. It is prompted by my recognized role as one of the nation's leading test pilots who has spent hundreds of hours behind General Electric engines in the most hazardous flying circumstances as the first American test pilot assigned to this country's first operational jet aircraft. The airplane was the Lockheed F-80 and the engine was the General Electric I-40. In case your G.E. experience does not include this particular jet, on March 20, 1945, I was almost killed in this airplane when the turbine disc disintegrated in flight shattering the rear fuselage with the loss of the tail assembly and complete loss of aircraft control.

I spent many painful months in the hospital recuperating from a fractured lower spine and only by the providence of God was my life spared.

During this period, General Electric employees, in whom I had great confidence, acknowledged to me that G.E. had experienced this same type of failure with this engine at your jet engine facility at Lynn, Mass., but had not seen fit to advise Lockheed up until that time. Subsequently, two other great American aviators, Test Pilot Milo Burcham and World War II Ace, Major Richard Bong, met untimely and apparently needless death behind the G.E. I-40 engine due to faulty over-speed governor operation.

But we live in a close community in aviation, a community which works together and, if necessary, suffers together. Thus, it was without hesitation that I straddled the G.E. J-79 engine in our Lockheed F-104 Starfighter series. Suffice it to say there was plenty of opportunity to remember my earlier experience with the G. E. I-40 engine. This engine kept the Starfighter program in jeopardy throughout its early life, but not only did we support G.E., not blabbing our problems with your product, we lent you both technical and moral support in correcting your problems.

This is the environment in which we at Lockheed continued to work with G.E. as a partner in those areas where our skills best complement one another...hopefully, without fear or favor. I am obviously not an expert on G.E.'s engine business, but I would hazard a guess that through the C-5 transport, the S-3A ASW aircraft and the AH-56, over and above the F-104 program itself, we are the largest user of G. E. engines in the world.

When we chose the Rolls Royce R.B.-211 engine for the Lockheed TriStar, we did not do it from weakness, but rather from strength. No one is more familiar with G. E. engines than Lockheed, but the Rolls-Royce commercial experience so overshadows G.E. experience that there was no room for serious contest. As a pilot with long experience behind General Electric engines, I am confident you will ultimately produce a fine commercial engine. If that should happen in 1971 or 1972, it will be in contradiction of the experience cycle of all other complex technical equipment in the history of aviation....whatever your experience with the CF-6 engine....and I wish you nothing but the best.

But, as a man who stood behind General Electric products when there was little cause to do so, and as part of a company which did the same, I condemn you and the General Electric Company for the crass manner in which you have operated in the matter of the proposed Lockheed loan guarantee. Despite my natural tendency to support big business, your transparent lack of good faith is disheartening to me personally and a disservice to General Electric and its thousands of stockholders.

Yours truly,
A .W. (Tony) LeVier

cc: President Richard M. Nixon
House and Senate Banking Committee members
General Electric Company Board members

The following ninety days were filled with negotiations between Lockheed, Rolls Royce, the airline customers, members of the British government, and personnel from DOD. Lockheed's bankers monitored all developments. The financial outlay to Lockheed had now reached the $400 million level. On 11 May the company and Rolls Royce signed a conditional agreement covering the development, production, and in service support of the RB-211 engine. One condition for making this contract final was the absolute requirement for Lockheed to obtain sufficient financing and continue the L-1011 program.

Delays increased the monetary needs of the company. A new finance program was established; an added $150 million, plus a contingency of $100 million was the requirement. No such financing was available from the banks without U.S. government backing in form of a loan guarantee, plus increased financial participation by the airlines. The airline additional participation was conditionally arranged pending the formal guarantee by the government of the $250 million portion of the credit line to Lockheed. Thus, government guarantee was the key to activating new credit agreements and the interrelated agreements with Rolls-Royce and the airlines.

On 7 June 1971, Dan Haughton, Lockheed Chairman of the Board, presented to the Senate Banking committee, detailed information on the need for a US government loan guarantee for the L-1011 program.(1) The report specified the events leading to the present state of the commercial and government contracts. Included in the presentation were cataloged interlocking effects that failure to provide the loan guarantee would generate—principally a loss of 17,800 jobs at Lockheed and 16,000 by supplier's personnel. Other facts presented included the following which also describes effects of the probability of Lockheed bankruptcy.

a) Loss of the L-1011 production would eliminate 33,000 direct jobs and an equal number of indirect labor positions in the commercial sector.
b) Close to $1.4 billion invested by Lockheed, 4,000 suppliers, and airlines could not be recovered since over 90% is applicable to only L-1011 requirements. Salvage value of this program investment was stated at two cents on the dollar.
c) The impact of lost federal income tax was conservatively estimated at $68 million. Declaration of company, airline, and bank losses will increase tax loss to the government.
d) Lockheed bankruptcy would adversely affect continuation of DOD programs.
e) In a letter to Congressman Moorhead, Controller General E.B.Statts pointed out, "Trustees of assets in bankruptcies have an obligation to all creditors-not only the U.S. government. There is serious question if the company would be required to perform contracts made in 1965 and 1966 when lower prices prevailed. Terminations would substantially increase government costs if the C-5 program continues."

Haughton's presentation did stress positive effects that the loan guarantee would generate. The preservation of over 33,000 direct jobs was the prime matter. And, the preservation of competition in that segment of the airline market would preserve a lower cost trijet to current and future customers. Conservative studies by airline and company analysts forecast a minimum 150% increase in passengers by the 1980's. Thus, a market for 775 basic, intermediate range trijets was considered the current market potential. (27)

Nor did he neglect the broader questions of how and why the corporation chose to compete in the commercial market when engaged in substantial military contracts. In essence, he pointed to program results that marked the company's technical superiority earned through taking on complex technical challenges. Successes such as the Polaris missile program, and the SR-71 Mach 3 aircraft generated internal fortitude to accept the tasks of the C-5A development and production, as well as the opportunities and challenges of the L-1011 TriStar. The admission to less than perfect management was discussed, but not offered in excuse. It was paired with parallel data placing 65 to 70 % of the ballooning cost factors for programs administered since 1967 as outside the direct control of the company. Assistant Secretary of the Air Force, P.N. Whittaker's testimony to the U.S. Armed Services Committee on 3 June 1971 stated the following,

a) "The TPPC contract considered the C-5 a state-of-the-art airplane, IT IS NOT.
b) To meet weight guarantees expensive raw material was required - high strength steel, titanium, beryllium, and fiberglass - in total weight approaching 65,000 pounds.
c) Cost growth was inherent in those problems associated with sheer size.
d) It is patently unfair to label as 'poor management' a cost growth 70% of which is outside the management of either contracting party." (28)

The bulk of the information in this 39 page report was a detailed analysis of these factors. Haughton's presentation stressed the decisive nature of the loan guarantee and provided a plan for use of the funding. It forecast first draw the third quarter of 1971, peak borrowing in the fourth quarter of 1972; the entire amount would be repaid by December 1974.

The Lockheed effort beginning in early1969 to conserve cash was a series of strong measures.

—Capital expenditures were reduced.
—Selective sale of corporate assets.
—Deactivated certain facilities, reduced size of others.
—Reduced inventories, cut overtime, reduced new business expense.
—Furloughed 8,000 L-1011 employees for 90 days plus.
—Reduced executive pay 12%; set a moratorium on employee pay increases.
—Deferred all promotional pay increases.
—Reduced total corporate manpower. By May 1971 it was 70,500, lowest in 10 year period.

Legislation, called the Emergency Loan Guarantee Act, was brought to the floor for vote in August of 1971. The bill passed the House of Representatives by a 192 to 182 margin. In the U.S. Senate the margin was 49 for and 48 against. The deciding vote was cast by first term Senator, M. Cook of Kentucky. (29)

Thirteen months later, on the first "anniversary" of the passage of this highly contested legislation, D.J. Haughton reported the following information in a corporate memo.

Representatives from both the U.S and British governments, twenty four lending banks, five airlines, Rolls Royce, Lockheed and members of the Treasury and Emergency Loan Guarantee board met to review Lockheed's status. These accomplishments were noted.

a) Current jobs intact; 16,000 at CALAC; 15,000 suppliers, and 30,000 in England.
b) The U.S. Treasury has a $2 million profit. The Emergency Loan Guarantee Board report of 31 July 1972 states fees received equal $2,076,276; expenses $131,480; leaving earnings equal to $1,944,796.
c) The DOD had uninterrupted delivery of Poseidon missles, Agena space vehicles, P-3 Orion airplanes, C-5 Galaxies, C-130 Hercules and other products.
d) Customer airlines saved $246 million in prepayments. Eastern and TWA have L-1011 airplanes in service.
e) For uncommitted airlines, a choice of airplanes exists.
f) For stockholders, a return to profitability; for suppliers avoidance of a $354 million loss.

Six additional subjects were covered. This summary statement by the ELGB was included.

"The board determined, on the basis of the record before it, that it could not find need for a guarantee that was the result of the failure on the part of management to exercise reasonable business prudence and that under the Act IT IS NOT necessary for the ELGB to require management changes."

The loan guarantee netted the federal government over twenty six million dollars in Lockheed-paid fees and five million in interest earned on these payments.

THE C-5A STRUCTURAL WING DESIGN AND REDESIGN

The C-5As had the inner, mid, and outer wing boxes replaced after thirteen years of flight operations.

Early in the operational life of the C-5A, the static strength of the wing, as well as its fatigue characteristics, were determined to be considerably below target values. Since flight testing and production were simultaneously being carried forward, this revelation resulted in substantial program changes for both Lockheed and USAF. However, those aircraft in flight status for test purposes or operational missions were flown in complete safety. All airplanes in flight status were limited to 80% of design air-load limits until all ground and flight test events have proven the design criteria.

The initial contract for design, development, and production of the C-5A was awarded to Lockheed-Georgia Company in October 1965. A two month contractural revision period delayed effective initiation of these processes until December 1965. In mid 1966, a potential over specification weight condition was identified to the Systems Program Office(SPO). Changes to the aerodynamic details of the wing, fuselage and empennage required the addition of 14,000 pounds to the weight of the empty airplane. These changes, verified by extensive wind tunnel testing, involved changes to the wing to fuselage fairings, wing leading edge flap attach fairings, and the engine pylon shape and its wing attach points.

The operating weight empty, OWE, was one of the contractual guarantees.

The requests for an increase in the airplane empty weight were denied. For the remainder of 1966 potential solutions that might solve the weight impasse were proposed to the USAF for evaluation. These proposals included an increase in engine thrust, which was negotiated with General Electric, the engine manufacturer, to assure the capability to achieve the take-off, landing, and cruise performance guarantees. This specific proposal was refused at the Assistant Secretary of Defense level. Negotiations in this direction terminated when a "Cure Notice" was transmitted to Lockheed in February of 1967. Thus, a comprehensive weight reduction program was initiated by Lockheed as a response to this administrative/contractual maneuver.

In early 1967 a Division Advisory Group(DAG) ad hoc committee was established by USAF to review the C-5A technical development status. This committee's charter required assessment of the C-5A design and development progress and evaluation of Lockheed's weight reduction plan. Conclusions reported were:

1) The contractor is doing a competent, imaginative job.
2) Specified performance appears to be attainable though landing distance may be exceeded by 250 feet.
3) The empty weight will likely exceed the guarantee value by 4,000 pounds.
4) All areas require vigorous efforts to meet goals.
5) A final comment, "fatigue design appears to be based on sound engineering practice, careful attention to detail, and the use of the best possible materials, processes and joining techniques."

Meanwhile, structural tests were in progress or accomplished using a half dozen major full-scale elements of the airplane. The test device and its function is listed here:

a) X999 - A test airplane, was the primary static test article for verifying structural capability of the wing, fuselage and landing gear interface structure.

b) X993 - Used to accumulate accelerated wing fatigue data.

c) X995 - Nose gear tests.

d) X996 - Main gear tests.

e) X997 - Aft fuselage/empennage tests.

f) X998 - Wing/fuselage fatigue tests.

Simultaneous with this ground testing, flight testing of the airplane began in June 1968. A total of eight airplanes were involved in flight test demonstrations. All aircraft were limited to 80 percent of design limits until ground tests are completed, a standard industry practice, and since it was rigidly enforced, all C-5A airplanes, whether in flight test or initial operational deployment, were flown under completely safe conditions.

In July of 1969 the wing of the static test article, X-999, failed at 125 percent of the design limit load. This critical event occurred at 25% below the target (required) value. The aircraft in flight status were essentially unaffected since they were operating at a flight limit which is at 80% of the design specification, providing a margin of safety greater than the required 1.5 value. This adverse test result prompted the formation of a Scientific Advisory Board (SAB) ad hoc committee by USAF to analyze test results and conclusions, review proposed wing beefup plans and recommend action. In their report dated 16 June 1970 the following observations and recommendations were made:

a) Lockheed employed higher stress levels in critical wing structure. To achieve the fatigue life requirements the quality of design detail and manufacturing application must be significantly better than has been achieved in the past.
b) Deficiencies uncovered in wing static strength by failure of the full scale static test article, X-999, resulted in modifications which are validated on component tests, but the full scale tests must be completed to validate adequacy of the wing fix. (30)

Five additional recommendations were included in the report.

a) Employ a Passive Air Load Alleviation System (PALS) in the wing. This is fundamentally an up-rigging of both ailerons which shifts air loads inboard toward the wing root, thus reducing wing bending stresses.
b) Accelerate the static and fatigue testing by combining loads conditions, reducing the test combinations and increasing the time between inspections.
c) Begin a second fatigue test on another wing as soon as possible, testing at the highest possible cyclic rate.
d) Begin an airload monitoring program on each aircraft immediately.
e) Accelerate parametric analysis of fatigue damage due to usage patterns.

Modifications to the wing were completed and addition tests were accomplished on the full scale ground test article, X999. The follow-on testing experienced a second failure on 13 March 1971 at a 126 % of design limit load. This test result essentially established the wing strength limitations defining a reduced aircraft weight-payload capability.

In December 1971 the Secretary of the Air Force directed the formation of an Independent Structural Review Team (IRT) to provide in-depth review of the C-5A program and to recommend solutions. This team consisted of approximately 100 experts from the United States and British aerospace industry, academic institutions, and the government, who spent fifteen months at the Lockheed-Georgia Company completing their analyses. As a prelude to the intense scrutiny which was mandated by this review of every facet of the C-5 program, one of the chartering remarks for the IRT is worth repeating. "USAF is committed to revise utilization and cargo limits to enhance wing life until the complete reassessment can be made of structural repair options."

The IRT report issued in early 1973 concluded that the C-5A, except for the wing, could meet the 30,000 hour service life goal. Wing life was estimated to be 8,000 flight hours at this point. Several months later the estimate was further reduced to 7,100 flight hours using new engineering analytic methods. The service life of the wing could be extended by various combinations of air load alleviation systems, fastener changes or design modification. The IRT report presented nine specific recommendations. This, then, provided USAF with options for design modifications and changes in operational procedures to extend wing life of the C-5A. In this report the feasibility of major wing modification, replacement of the center and inner wing boxes, was presented as one of the options. The near term solution involved alternate fuel sequencing routines for the flight crews and incorporation of the PLDCS system. An urgent concurrent recommendation was made for an active (full-time) air-load alleviation system(ALDCS). This was accomplished. The long term plan defined a modification/rework schedule for the wing. This plan, designated plan "H", was selected as a baseline which was updated by the results of additional testing and analysis.

A Scientific Advisory Board(SAB) ad hoc committee was established by Lt. General James T. Stewart, Commander of the Aeronautical Systems Division (ASD). In his remarks at its opening session on 13 December 1971, he emphasized that the C-5 is indeed a critical element in the force structure, a national asset, and it is absolutely essential that any doubts or concerns that may exist with respect to the structural integrity of the aircraft be resolved. In April 1973, the Scientific Advisory Board (SAB), whose charter required monitoring and guiding the efforts of the IRT, reported the following to Commander of the Aeronautical systems Division.

a) The SAB agrees with all major aspects of the IRT findings.
b) The SAB concludes the fundamental problem plaguing the successful completion of static and fatigue tests and jepordizing service life is the high stress levels used in design.

Thus the IRT committee's work was the lead into a four year effort analyzing the economics, the manufacturing methods, improving structural analysis techniques, and a meticulous revision of operational utilization of the airplane. The data bank of information on the C-5 airplanes was very closely monitored in the Structural Information Enhancement program (SIEP). One of the keys in this process was a strong recommendation from the USAF Scientific Advisory Board (SAB) meeting in September 1977. They recommended the use of fracture mechanics engineering techniques, at the earliest possible time, as the basis for determining the useful life of the pre-modification C-5A wings, preempting the use of fatigue analysis as the basis for determining useful structural life.

Wing Modification

The wing modification program consisted of four phases: design, test, fabrication, and installation. The design phase contract was awarded in January 1976. The test phase, authorized in January of 1977, consisted of the manufacture of two ship sets of wings—one to be installed on the fatigue test article and the other on the flight test airplane—for dedicated testing. The cyclic loading of the fatigue test article operated through two equivalent lifetimes, which is 60,000 cyclic test hours (CTH).

The flight test airplane, C-5A (#68-0214) was delivered to Dover AFB on 16 January 1981. It was flown by aircrews from the 436th Military Airlift Wing for the remainder of the year accumulating over 1,000 flight hours. The wing performed satisfactorily. Flight testing was continued through December of 1982. Fuel leaks developed around riser clips in a number of the fuel tanks. Each wing has six integral fuel tanks. The contractor changed sealant application methods which eliminated the problem. Fatigue strength of the prototype wing met design requirements achieving 60,000 CTH. Follow-on damage tolerance testing was initiated when the test managers had artificially introduced sixty(60) flaws to check structure stability. At 75,000 CTH inspection revealed only slight growth of flaw size. An additional ten flaws were cut in the test article and testing continued through 105,000 CTH with satisfactory test results. Upon completion of the 105,000 CTH, the wing was subjected to an additional series of tests to determine residual operating capability with severely damaged wing structure. These tests were concluded satisfactorily when the wing achieved 100% design limit load with the severely damaged structure. (31)

The wing modification involved replacing the wing boxes tip-to-tip. The leading and trailing edge structure along with all control surfaces, ailerons, spoilers, flaps and slats were salvaged from the old wings and reinstalled on the new wings. The ALDCS system has been retained on all C-5A and C-5B aircraft. Seventy seven C-5As were modified. Five airplanes have been attrited; two due to ground fires, three due to operational accidents. The first production modification began in January 1982 with airplane 0012. The final modification was completed in May 1987 on ship 0081. (32)

C-5A with T-38 chase.
Author Flying Chase Aircraft

CHAPTER 4

THIS CHAPTER EXPLAINS WHAT HAPPENED TO "KISS"

GROUND AND FLIGHT TESTS

The numerous procedures used to qualify the C-5 Galaxy for daily operational use can be described in two simple, definitive categories: ground tests and flight tests. These labels, however, are where the simplicity ended. Complete verification of all specifications required extensive expansion of testing capacity. Full-scale mockups of all basic airplane systems were built on mechanical devices labeled simulators. A simulator is a working mechanism. Its function is to duplicate system function, generally in a linear manner, allowing very close observation of all components under test. It also permits rapid completion of the cyclic requirements of the components under test. Dimensionally exact simulators were fabricated to test hydraulics, fuel, electrical, pneumatic, and landing gear systems. Avionics systems and automatic flight controllers were also evaluated in ground and flight laboratories.

Airborne tests were completed on three specially configured airplanes. The first was a B-52, modified with one TF-39-1A engine installed on an inboard pylon. The second airplane, a C-141A, functioned as an airborne test laboratory for C-5 communication, navigation, and radar systems. A third airplane, the Calspan Total-In-Flight simulator, was utilized to evaluate the stability augmentation (SAS) system computers for the flight control system. Additionally, eight of the first twelve C-5A aircraft produced were dedicated to contractor and customer flight tests.

The test program was completed with the predetermined policy that safety was preeminent over any other consideration. Both Lockheed and USAF management were determined to make certain every test procedure would be as safe as human ingenuity could make them. In flight testing Lockheed-Georgia established an enviable record. For a thirty-year span beginning in 1957 the company test record was spotless; no aircraft or aircrews were lost.

Category One (Cat-1) testing is that series of ground and flight-test events conducted exclusively by the contractor personnel. Category Two (Cat-2) tests were performed primarily by USAF personnel in an environment which duplicates operational realities the airplane will encounter. Tests in Cat-2 have a distinct bias toward mission completion under the "military order of business." Cat-1 tests included, but were not limited to, the following variety of ground and air tests:

a) Completion of the flight controls proof tests and the structural rigidity tests prior to the first flight
b) Ground vibration tests, which correlates estimated airplane structural vibration characteristics with theoretical analysis and wind tunnel flutter model results.
c) Flight testing to certify freedom from flutter possibilities
d) 80% structural integrity flight tests.
e) 100% structural integrity flight tests.
f) Validate by demonstration the takeoff, climb, cruise, descent and landing performance at all critical gross weights.
g) Flight tests with one or two engines inoperative to define the limits of safe operation.
h) During flight, intentionally disable electrical and hydraulic systems to validate safe flight with normal controls partially disabled.
i) Results from all flight testing were used to develop crew procedures addressing every reasonable flight contingency, from the loss of thrust on two engines simultaneously, to a most extreme flight condition—a fuselage fire while cruising at high altitudes.

Ground testing of the Galaxy airframe, and essential primary and secondary systems, was accomplished through the use of three airplane-sized laboratory test specimens, designated X-997, X-998, and X-999. These were built along with six sub-system peculiar test articles designated X-991 through X-996. Load bearing capacity, component strength, endurance characteristics demonstrated by cyclic tests, and fatigue life were the principal areas of structural testing. Demonstrating the airplane's ability to operate safely in the full spectrum of adverse weather extremes, which are the military operations environment, was the objective which drove the scope of these tests. The three largest laboratory test specimens are described below with the specific test objectives noted.

(X-999)—This full scale fuselage and wing test specimen was built to accommodate static loads tests of the airplane. Static loads of increasing value were progressively applied, through a system of hydraulic jacks, to the wings and the backup fuselage structure until ultimate load levels were reached.

(X-998)—a second full scale wing and fuselage test article was used to achieve complete wing and fuselage fatigue tests and demonstrate fuselage pressurization capability. These tests could not have been accomplished merely by pumping air into the test fuselage specimen. A component failure under these conditions would have the effect of suddenly releasing trapped energy, a bomb-like effect. This hazard was avoided by immersing the complete fuselage in a tank of water. This 250 foot long tank was another laboratory tool. Since water is virtually incompressible, testing of suspect structure became safer.

(X-997)—This test article was used to test the aft fuselage and empennage structure for integrity under the expected flight and ground loads, as well as prove fatigue life characteristics. To complete cyclic (fatigue) tests on the nose and main landing gear, separate test articles with the necessary backup structure were built. A second full-scale wing and barrel test article, designated X-993, was the accelerated fatigue test device. Wing structural integrity tests which included fatigue cyclic tests, crack growth tests, and fail safe tests with severed wing surface panels under limit loads were performed on a laboratory article designated X-991.

Additional ground test mechanisms were built, such as a full-scale cockpit to test the windscreen's ability to absorb bird impact at speeds up to 285 miles per hour. Actual chicken carcasses (unfrozen) were fired at the windscreen by a device known as the "rooster booster". A prime ground test device was the full-scale flight controls test unit. This was constructed on an "I" beam support structure. Thus, a means was provided for cyclic testing of all the flight controls and their unique actuating components. The primary and secondary controllers were cycled thousands of times to assure integrity of these basic components. The "I" beam structure, referred to as the "iron bird", was housed in one bay of the B-4 building, an existing hanger adjacent to the principal production facility, identified by the simple label, the B-1 building.

Housing of the special lab devices to accomplish the structural tests required construction of the L-10 Engineering Test Center, modification of the existing B-25 flight test hanger and building supplementary warehousing facilities designated L-11 and L-12 buildings. The largest of these buildings, the L-10 hanger, has many distinctive characteristics. The L-10 has floor space to house four C-5 airplanes simultaneously, in addition to providing a central, high rise, office complex of five stories. This entire structure under one roof is easily identified by the exposed steel truss structure supporting the clear span of 490 feet in each of the two hanger bays. The west hanger bay was the location of the majority of the cyclic and fatigue testing.

Figure one (located on page 48) is a composite schedule illustrating the integration of flight test events. The highlight on figure one is the October 1965 forecast for the first flight of the C-5 Galaxy to occur in June 1968. Typical ground testing required to be completed prior to first flight were the flight control system cyclic and proof tests, the structural rigidity tests, and the 80% static loads tests of primary wing, fuselage and empennage structure. Operation of the basic flight control system (on the iron bird) through 10,000 cycles were necessary; this represented ten percent of the total cyclic program requirements. Landing gear structural tests on static loads article X-999, were 90% complete by the first flight date.

The bulk of Cat I and Cat II flight-testing was accomplished on C-5 airplanes 0001 through 0008. The assigned tests and estimated flight hours were as follows.

Test Description	Flight Hours
SHIP 0001	
Preliminary flight evaluation	24
Stalls, air-load survey, flutter tests	200
Flying qualities, CAT 1 and CAT 2	308
Automatic flight controls evaluation	202

Figure 1: Composite Flight Test Schedule

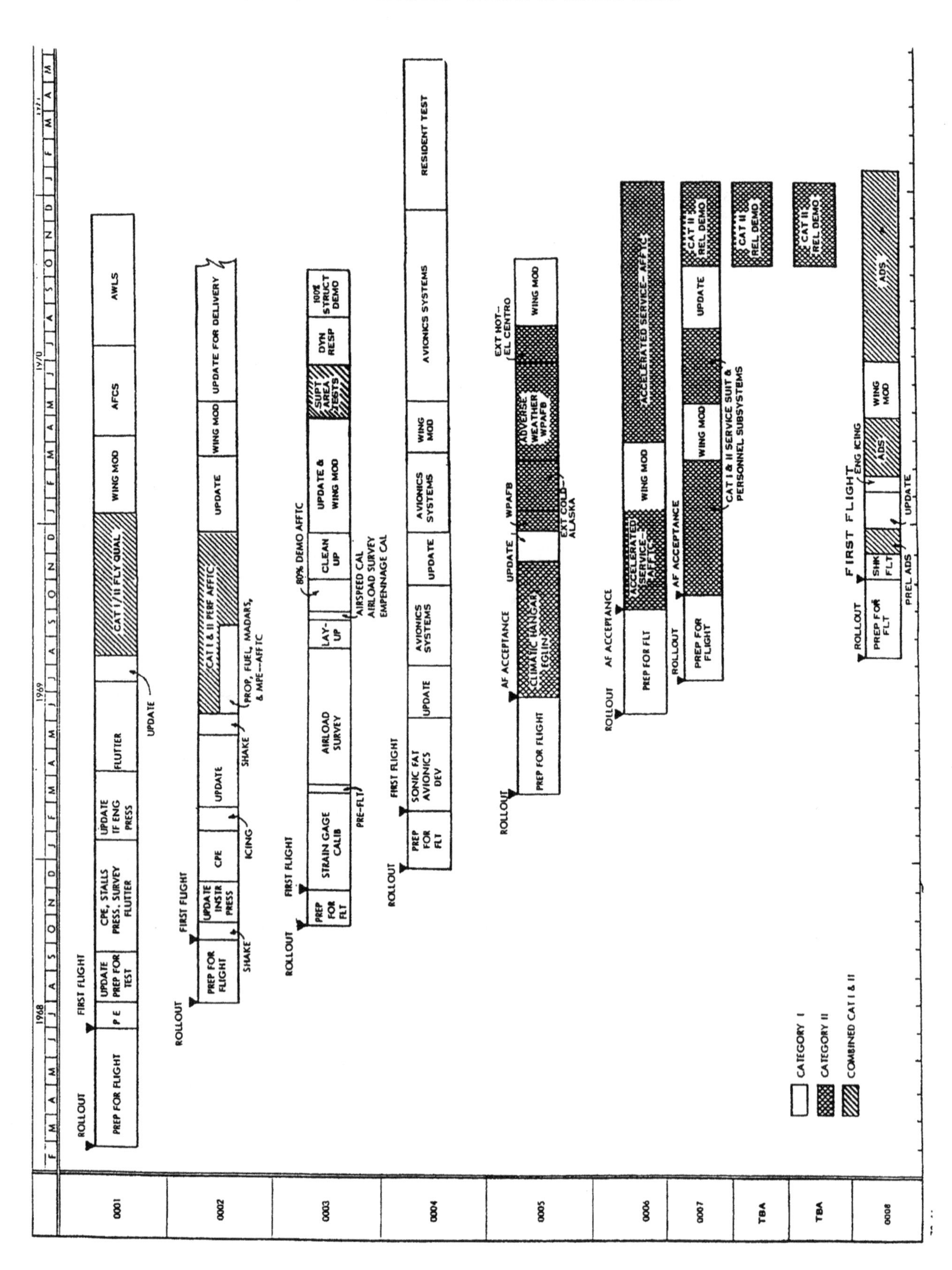

Test Description	Flight Hours
SHIP 0002	
Initial systems, airworthiness evaluation	31
Icing tests, engine tests	15
Fuel system tests, MADAR, engine tests	125
CAT 1 and CAT II performance tests	240
SHIP 0003	
Shakedown flight, evaluate instrumentation	6
Air-loads survey	135
80% Structural survey demonstration	70
CAT 1 and CAT II support area demos	90
100% air-load structural demonstration	90
Dynamic response testing	110
SHIP 0004	
Shakedown flight, sonic fatigue measurement, avionics systems development	124
Integrate and demo avionics systems	282
Mechanical and avionics subsystems	220
SHIP 0005	
Air worthiness, Engine qualification	36
In-flight refueling tests	20
Airspeed calibration, performance tests	77
CAT 2 adverse weather testing conducted by Wright Patterson AFB 4950th test group.	–
SHIP 0006	
CAT 2 accelerated service test aircraft. Tests completion date: 1/31/1971	2700
SHIP 0007	
CAT 2 Service suitability, primarily ground demonstrations.	–
SHIP 0008	
Preliminary aerial delivery(ADS) tests	90
Airframe and engine icing tests.	30
CAT 1 and CAT 2 ADS; Electromagnetic interference(EMI) tests.	24

The earliest forecast of total flight hours required by the contractor was 3000 hours. As flight-testing progressed a better understanding of the airplane characteristics and on board subsystems dictated an expanded test schedule. An additional 455 hours of flight-testing were added to fulfill all requirements.

Proper flight-testing always requires dedicated instrumentation and specialized equipment in the cockpit of test aircraft. Due to the aircraft mission and the size of the airplane, instrumentation and recording equipment were a major set of requirements on their own. In the case of the C-5, appropriate sensors were placed throughout the fuselage, on the wings and empennage to measure structural strains/loads and control surface positions. In addition, unique devices were designed to assist in the evaluation of aircraft damping characteristics during flight flutter investigations. Comprehensive onboard automatic recording equipment, as well as ground telemetry devices, were utilized extensively on the first three test airplanes. Also, T-33 and T-38 jet trainers were extensively used as chase airplanes. Routinely each chase airplane carried a cameraman to photograph those system operations which cannot otherwise be monitored. The following list emphasizes the complexity of the test equipment utilized to acquire flight results.

Test Equipment

	AIRPLANE ASSIGNED	0001	0002	0003	0004	0008
1.	Wheel well fire extinguishers	x	x	x		
2.	Spin chute	x	x	x		
3.	In flight egress equipment	x	x	x		
4.	Wheel visual inspection unit	x	x	x		
5.	Nose boom(airspeed, altitude)	x	x	x		
6.	Water ballast system	x	x	x		
7.	Aft fuselage skags (over rotation protection)	x	x	x		
8.	Takeoff and land camera	x	x			
9.	Free air temperature probe	x	x	x		
10.	Wiring, plumbing tests	x	x	x	x	x
11.	Ground spotting camera		x			
12.	Trailing air pressure cone	x	x	x		
13.	Test cockpit instrument panel	x	x	x		
14.	Wing tip and horizontal stabilizer tip flutter excitation vanes			x		
15.	Telemetry system	x	x	x	x	x
16.	Special auto flight control units	x	x	x	x	

Dobbins Air Force Base in Marietta, Georgia was the principal base of operations for flight testing the C-5A Galaxy. Meticulous flight tests to define flying qualities, measure air loads, validate structural integrity, evaluate engine thrust characteristics, and certify that all on-board systems performed to specification followed in detailed succession. The avionics systems, which included communication, navigation and automatic flight control systems, were the focus of day and night long-term flight investigations. Flying quality tests have reasonably straightforward objectives. They seek to prove the manufacturer's claim that the airplane was easy to fly, and safe to maneuver to the extremes of its flight envelope. Complexities of this plain vanilla task arose in the gathering and recording of accurate test data, which would satisfy the host of maneuverability requirements. One of the flying quality specifications involved demonstrating that the airplane is capable of achieving eight degrees of roll displacement in the first second of a turning maneuver. This relates to a sidestep maneuver necessary for runway alignment after descending below the clouds (breakout of the overcast) which are only two hundred feet above ground. This flight requirement was a major element driving the design of the flight control system, which in the final configuration of the lateral control system, is a combination of ailerons and wing spoilers with wing flap position mechanically limiting spoiler deflection angles.

The reader may remember that earlier in Chapter 3 the cubed/fifth power law was discussed and its effect in understanding the maneuvering characteristics of the Galaxy. Since the C-5A is roughly twice as heavy as is the C-141, they differ by a scale factor of the cube root of two, which is 1.26. Raising this number to the fifth power equals 3.18; so all other variables being equal, the C-5A size alone makes it 1.59 times (3.18/2) more difficult to do any maneuver than an airplane the size of a C-141. The Air Force ultimately relaxed the roll angle value to 6.5 degrees.

The Galaxy has a Cooper-Harper (CH) flying quality rating of 1.5. This superior characteristic continues to impress pilots and flight crews even today, thirty-three years after first flight. Structural testing involved very precise flight maneuvers at the extremes of the airspeed/air-load envelope at all critical operational gross weights. These are standard fare for cargo airplanes. However, the flutter test involved a daunting new set of complications for an airplane this size. To prove that the Galaxy flight controls and adjacent structure were free of flutter tendencies required specially designed devices to force wing tip oscillations. Paddles, hydraulically operated from the onboard instrumentation panel, that would deflect the air-stream, were installed on the tips of the wings and the horizontal stabilizer. Under the necessary test conditions these paddles were cycled in the free air, individually, simultaneously, or asymmetrically as testing required. The measured response of the wingtip was to oscillate as air loading radically changed. After a few seconds of this agonizing wing load redistribution, these devices were nulled and natural wing damping was duly recorded—with satisfactory results. The photo to the right illustrates the specially designed devices used to produce wing oscillations during flight tests.

Engine testing was assigned to ships 0002 and 0005, but, in practice, all test airplanes contributed to information database. Flight tests followed more than twenty-four months of extensive ground tests at GE in Evendale, Ohio, and Lockheed in Marietta, Ga. In addition the Air Force provided a B-52 for use as a flying test stand. One TF-39 turbo fan engine was installed in place of two of its inboard engines. This airplane/ engine accumulated over 300 flight hours prior to first flight of the C-5A. Engine testing involved charting the characteristics during all flight phases; define the air start envelope, ie, airspeed and altitude limits; and conduct reverse thrust operations in the air and on the ground. (In 1985 ground tests were conducted using engine reverse thrust for taxiing the C-5 airplane rearward.) At take-off power the high thrust levels had some "side effects" that were initially very vexing. Until brake release for takeoff roll, the airframe response was a mild shaking motion. In addition, the gale like airflow behind the airplane had the expected consequences on loose terrain, equipment or other objects within 500 feet of the engine exhaust.

Wing mounted hydraulically activated paddles to oscillate the wing.

The C-5A avionics systems initially included a radar system capability designed to direct flight in a defensive maneuver known as terrain following. This is navigating a desired track at a minimum designated ground clearance height. The testing occurred over level terrain (farmland), rolling hills(grazing land) and portions of the Appalachian chain (north Georgia and southern Tennessee) at a clearance height of 500 to 1000 feet. This close to the ground maneuvering of this giant airplane on a variety of tracks produced several non-scientific test results. Repetitive appearances of the C-5A over these test tracks in a variety of weather generated a hail of irate citizen responses, vocal and written, to anyone at Lockheed answering the phone or opening unsolicited mail.

Several test bases were used during the many months of flight-testing. Prime among these were Edwards AFB and El Centro Naval Air Station in California, Eglin AFB in Florida, Eilson AFB in Alaska, and Pope AFB in North Carolina. Take off and landing performance tests were conducted at Edwards AFB environmental tests in the cold hanger at Eglin AFB were followed by winter weather flights from Alaskan airfields. Dramatic, essential testing of the aerial delivery capabilities were flown from Pope AFB. Also, in the course of automatic flight control demonstrations, many hours were flown in the ILS pattern at Robbins AFB,GA, but the flights normally initiated and ended at Dobbins AFB. Airplane deliveries to the USAF began in December of 1969. The first airplane to be delivered, C-5A 0009, was flown to the transitional training unit at Altus AFB, Oklahoma on 17 December of that year. Appropriate ceremonies were part of this significant event.

General Fergueson, Commander, Air Force Systems Command, accepted the airplane; General J Catton, Commander of MAC, flew the airplane to Altus AFB. The initial five deliveries were to Altus AFB. Later deliveries alternated between Charleston, Dover, and Travis AFB.

Off paved runway testing at Edwards AFB.

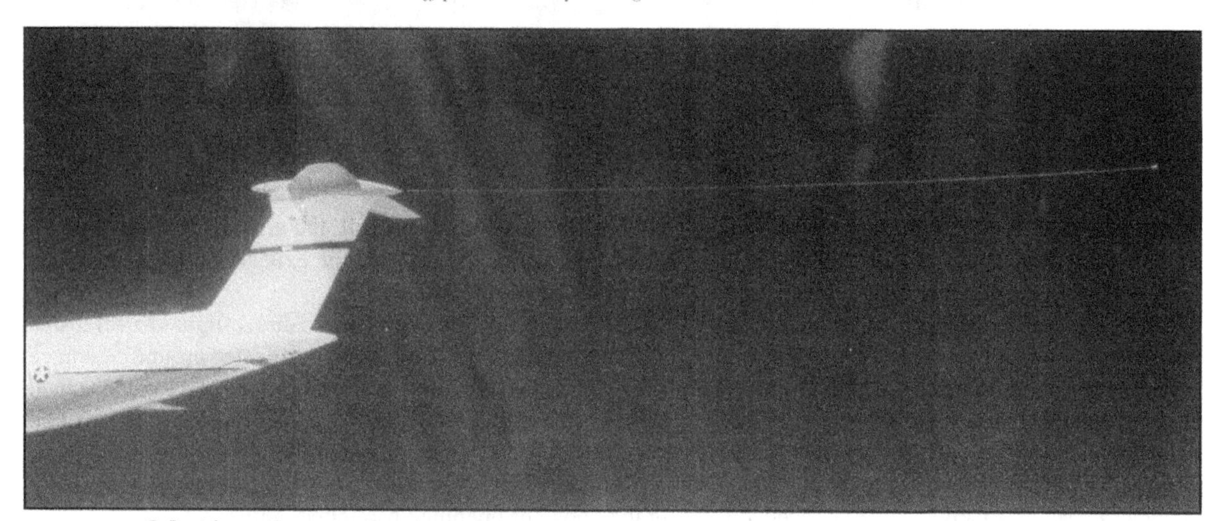

C-5 with a tail cone trailing from its vertical stabilizer – a flight test device for airspeed calibration.

Radome off tests at Edwards AFB. G. Gray at the controls. USAF photo.

C-5A shadow, north Georgia terrain following test, 500 feet AGL.

TF-39 engine test stand. Lockheed photo.

Pre-flight briefing for the press, three days before the first flight. From left: Edwards, Hensleigh, Schiele, Mittendorf, and Sullivan.

Chapter 5

"Liftoff at 0747"

Mitt Mittendorf

First Flight for the C-5A Galaxy

Airplane first flights are always cause for special interest. There is excitement for a planned first-ever event. Despite the fact that the C-5A represented significant advancement in aerodynamics, structures, clever systems designs, and dozens of other spectacular advancements, the one characteristic that overrode all of this was its size.

This, coupled with the fact the airplane would be flown on its maiden flight by a five man crew, begged explanation. The world's biggest airplane, at that time, was being flown by the absolute minimum crew of five. Leo Sullivan and Walt Hensleigh, the Lockheed pilots, Lt. Col. Joe Schiele, US Air Force pilot, Jerry Edwards, flight engineer, and Ernst (Mit) Mittendorf, the flight test engineer, were the select few.

Minimum crew size was, and is, an inflexible rule for all competent flight organizations. The corollary to this first flight requirement dictates that flight checks of basic systems in their normal functioning mode be accomplished rapidly and accurately. Flight is restricted to remainwithin a reasonable distance of home base. All these obvious, yet absolutely essential elements, impact how, when, where, and how soon the airplane will be flown off the runway. Weather is definitely a determining factor.

How does one qualify to fly, for the first time ever, an airplane of this immense size? In addition to other considerations the C-5 is powered by newly designed engines. Thus, the crew had new airframe and new engines, a dual combination of untried technology that required full focus, and then some, to complete the mission.

There were no aircraft even remotely similar to the Galaxy in flight status that could be utilized as 'training devices". There were however a goodly amount of speculative suggestions from both the SPO and senior Lockheed management for methods to fill this relative unknown on the checklist itemizing proper preparation.

The commentary below reflects vivid imagery that is still prominently recalled more than thirty years after this historic flight.

Leo J. Sullivan, Chief Engineering Test Pilot from 1956 to 1968.

Q: What were the special concerns in the minds of the USAF and the management as first flight date for the C-5A approached?

LEO: Specialized training in systems and rigorous understanding of estimated flight characteristics were the one area that the company managed with ease. Flight training, if one could call it that, was a different matter entirely. Flight in anything close to the size of the C-5 was considered a reasonable idea by some; mandatory by some in the USAF, especially Guy Townsend. Thus we spent a lot energy trying to do just that. Walt and Hank (Hank Dees was Walt's stand-in if he was in sick bay on the 30th of June) went to Edwards AFB to get some time in the B-70. But not all the equipment was working properly so the exposure was of questionable value.

Q: Did you get a turn at the B-70?

LEO: Yes, but once again, the size of the airplane was the only reasonable comparison. The flight control characteristics were grossly incompatible to the machine I would fly in June. I muttered a few "Hail Mary's" with the mental request that the C-5A be infinitely better. Prayer answered!

Q: What was the thinking in the requirement that you get in flight refueling (IFR) qualified prior to first flight?

LEO: This was a pretty arbitrary item thrown in by Townsend. So, about six weeks before the first takeoff I found myself in a B-52 doing in-flight refueling. Flying that big sucker makes one appreciate the excellent work our flight controls engineers accomplish on the Lockheed airplanes.

There is a need to understand that a pilot develops a sense of what he can handle on these new airplanes and he goes with it. Call it a gut reaction to the stuff the engineers are describing in their lectures. You always build on what you have worked with before. We had our own thoughts on what would help assure good results. I did some extra flying in a C-141, since it had a reasonable similarity of flight control and airplane response; then flew several very precise airport patterns in a C-130B to help fix in mind what the scene out the window should be in the landing pattern. The C-130B was the only airplane available the last few weeks before the 30 June flight.

There were months of formal and informal discussions on flight controls, engines, and the basic systems. The biggest plus was the crew going with you. Jerry Edwards did everything a flight engineer could possibly do to be knowledgeable about the systems, plus, he probably did a lot of flight line work without a union card. Pilots don't come any better than Walt and Joe Schiele. Mitt was with me on several other first flights; that doesn't happen if you leave your common sense at home.

Q: Making a first flight date scheduled three years earlier was a great feat; to do it on an airplane as big as the C-5 with all those complex systems must have raised some moments of doubt.

LEO: There were problems down to the wire for certain. Still everyone worked in a double-quick mode those last three months to assure success. We made everyone aware that we would not tolerate any hydraulic leaks, especially in the landing gear area. As it worked out, we did see several on those early taxi runs. But maintenance did the job, and we were perfect on the 30th of June.

Lt. Col. Joe Schiele, USAF, Pilot

JOE: It's obvious that Mitt's memory of the first flight events is much better than mine since he is referring to his personal log of the flight. There are a few thoughts that all this conversation brings out. I had considerable experience in "old Shaky", the C-124. It had a cockpit that was higher than either the C-133 or the C-141 cockpit. Those two prop airplanes were still in use then. So, getting used to the C-5 cockpit location was a snap for me. Where I had a real surprise was the ease of controlling the airplane in landing approach. The C-5A flies so slow that one feels that you can land the airplane on any spot on the runway that you wish and have a routine rollout.

Q: Didn't you feel a little awkward in the transition to touchdown?

JOE: Not at all, the cockpit visibility allows the pilot to progressively evaluate his advance on his intended touchdown point all through the maneuver without any strain. I never felt the need to stare at the radar altimeter, because it was natural to be heads up all the time, and this airplane did nothing to change that technique

Q: You had a few other first impressions I'm sure, what were they?

JOE: The flight control system was the biggest improvement over those other airplanes that I had flown. True, the C-5A had the power, it was much bigger; but the aircraft response to flight control inputs was amazing. The light stick forces coupled with the remarkable response was the real eye opener for me. In "old Shaky" for instance you had to muscle the bird around to fly it, sort of like going to the weight room, you know what's coming and you are prepared to work at it. In comparison, the C-5 pilot control force required never got close to the "work out level".

Q: Did you observe others during transition flying?

JOE: Oh, certainly. As you know, the first pilots to fly a new machine become "one landing IPs" (Instructor Pilots) and, as such, get to observe a lot of novel techniques for doing the job. Still, I never met anyone who had any difficulty in transitioning to the C-5A.

Q: Did you stay at Marietta very long after the first flight?

JOE: I flew on the first five flights, the airplane was put into rework, then, for about six weeks. I stayed in Marietta for some additional flying, and I left the program in March of 1969.

WALT HENSLEIGH, C-5A PROJECT PILOT

Q: Although the first flight of the C-5A is undoubtedly a memorable event, the pressures to make that very important flight date must have been very stressful. What was it like?

WALT: Certainly each one of us was more than a bit anxious as the flight began to be seen as a real probability, instead of just a note on the calendar. There were a number of very positive activities that helped keep us focused and confidant, however. The first and foremost was our essentially unlimited, open access to every level of the engineering staff to discuss our concerns and get answers to innumerable questions that came up. We really felt very privileged to get the quick-turnaround answers we needed and having that kind of cooperation across all organizational boundaries made our concerns about meeting our first flight target date quite manageable.

My particular assignment required that I work very closely with the Stability and Control and Flight Controls groups as they developed the configuration and systems to meet the extremely rigorous flying qualities requirements specified in the contract. We were making the first serious use of a piloted flight simulator to develop flying qualities in a large, complex airplane and we were breaking new ground every day. The hybrid, fixed-base simulator was rather primitive by today's standards, but we couldn't have come up with the entire flight control system, essentially in its final configuration on the first airplane and prior to the first flight, without that simulator.

Because it was so early in the development of sophisticated simulators, the simulation staff had to invent things as they went along and reliability was not really great. Sometimes it was a real challenge just to get the simulator up and running, load in the problem that was to be studied, do the required checks, get the pilot in the seat and actually get some useful data before something broke, burned out, or otherwise crashed! Hank Dees, Glenn Gray, Jesse Allen and I were the principal "worker bees" from the pilot organization for this effort, so by the time of the real first flight, I had already "flown" the airplane a lot in the simulator and felt like I had a good idea of what to expect.

Another thing that made anticipation of first flight a little less stressful was the redundancy that had been designed into the hydraulic and flight control systems. We had separated each of the usual control surfaces into two surfaces, each powered by a different hydraulic system and each hydraulic system could "cross-feed" power to another hydraulic system without any transfer of fluid. There was just no way for a failure, or even multiple failures, to cause us to lose control of the airplane. On the first flight of an all new airplane, that is a good thing to have on your side.

Q: The C-5 control wheel is limited to 60 degrees rotation either side of neutral. How do you like that?

WALT: It is a very good improvement and came out of the simulator-based development program I just described. We were all familiar with the C-141 and C-130 control wheel mechanical characteristics that required 90 degrees or more of rotation to get full throw on the ailerons. In turbulence or rapid maneuvering, that is a lot of work for the pilot. By reducing the wheel throw on the C-5 and at the same time hooking that into a control system that was a lot more powerful than either the C-141 or C-130, you get a much more maneuverable and better flying airplane at what results in a much lower pilot workload.

One other unique thing about the C-5 flight control system addresses the fact that, although it is a fully powered FCS, conventional control cables run all the way from the cockpit to the power servos next to each control surface and that is a long way. For the pitch axis, it is nearly 250 feet from the flight deck back and up the vertical stabilizer to the elevator servo. In a cable run that long, you can't avoid picking up a lot of friction from all the pulleys, fairleads and seals the cable runs through. To keep that from translating into high stick forces and slow control response, a cable assist servo was designed into the elevator system. It acts like another "power steering" unit by picking up the pilot's input at the yoke and multiplying his force input into the cable. We pilots had a big hand in tailoring the characteristics of this mechanical innovation.

Q: How did you feel about getting pre-first flight exposure to the B-70, which also had a cockpit quite high off the ground?

WALT: It was a novel experience and it was useful from that standpoint. But as far as preparing Hank Dees and myself for flying the C-5 is concerned, I think the value was probably overrated a bit. Even so, I have to say that it made a great entry in my logbook!

Q: How much did the elevated cockpit bother you?

WALT: Not one bit. Actually, I think most pilots had more trouble lugging their 25-pound parachutes up the access ladder than they did getting used to the "upper office". Every new cockpit has characteristics that the pilot has to adapt to quickly, but realize that this is a very big cockpit, the outside visibility is excellent, the flight and engine instrument panels were very readable and our test pilot group had a great deal of input into the design and functional layout of this area. As a result, this wasn't much of a transition concern for me, personally.

Q: So, was the first flight exhilarating, pleasant, or a "thank goodness it's over" relief?

WALT: Well, it most certainly was a great thrill, particularly since handling qualities were so impressively good in the crucial high drag configurations. With that said, though, everyone in the flight crew, myself included, was glad to get the first flight out of the way and into the history books. There was so much visibility and public attention focused on that flight, locally and nationally and we were anxious to get on with the real nitty-gritty flight test work. We knew that the first six months of the flight test program would be an unrelenting schedule of something new or expanded on every C-5 flight, expanding the flight envelope or evaluating one of the many areas of new capability that were inherent in this big machine. We had plenty of work to do, none of it routine and we were more than ready to get on with it.

Q: Did "split crew" flight testing result in any "interesting" situations?

WALT: Not as long as a reasonable amount of diplomacy was practiced. I found the USAF flight crew people well prepared to make the C-5 flight program productive and I never had a problem with any of them.

Q: Did you get out to Edwards AFB to do any flying?

WALT: I did some amount of flying off base, but my job promotion kept me fairly close to the home office most of the time. Still, I did get my licks in on some very interesting flights. As you remember, we had a water ballast system installed on a couple of the test airplanes to simulate varying cargo loads (and gross weights) and let us move the center of gravity around. The system of tanks was tied down to the cargo floor and could hold over 8,000 gallons of water. At the end of the flight, we normally would jettison that water on our way back home, to reduce landing loads and stress on the airplane. If we were above about 7,000 feet the jettisoned water would evaporate and none would get to the ground. At that end-of-the-test-card point on one of my flights, I was near the little Mississippi town I grew up in. I thought some extra rainfall might cool down the summer day and couldn't resist announcing my presence. I flew over the main street at an altitude that was, well, a little below the usual 8,000 to 10,000 feet and toggled the water jettison switch. I heard later that it was more of a localized flood for a couple of blocks along that street. Maybe they've forgiven and forgotten by now.

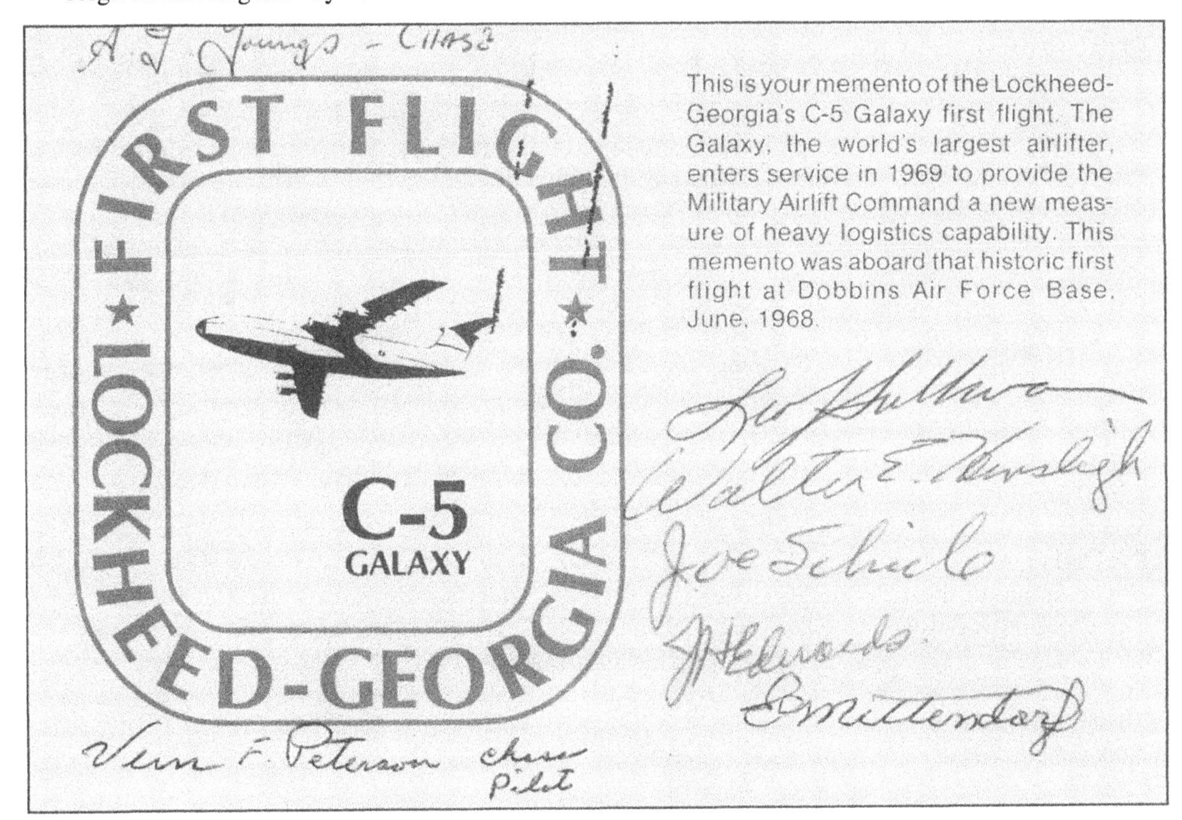

Q: There are obviously a lot of things about the C-5 that you like. What is the one facet or quality that you like most of all?

WALT: That's not as easy to answer as you might think. To have an airplane as large and potentially ungainly as the C-5 and still have it relatively easy to maneuver on the ground, plus have such exceptional handling qualities in the air, is really spectacular. And, that applies at all but the very highest gross weights, too. As proof of that, a number of C-5 engineering and production people, who were also low-time, limited experience pilots, were given a little "courtesy time" in the seat, flying the airplane on later more routine test flights at altitude, of course and none of them had any difficulty at all in flying the airplane and doing a pretty decent job of it. All of them were amazed that the C-5 was that easy to fly and not strikingly different from the light planes they were more familiar with.

If I'm really pressed for the C-5's single best characteristic, I think it is the ease with which the airplane handles and maneuvers in turbulence or in instrument conditions. In those areas, it absolutely has no peer.

Mitt Mittendorf, Flight Test Engineer

Q: You must have some strong recollections of the C-5 first flight.

MITT: I certainly do. The flight itself was a relief after the very hectic week preceding it. As you might suspect the world's biggest airplane drew a big crowd. From the first time the airplane taxied there was always a crowd, some of which were actually working on problems concerning the flight. Naturally the media was here in a pretty large numbers. They weren't too difficult to manage. It was different at the post flight, for sure. Tom May, the division president presided. We taxied on six consecutive days prior to flight to thoroughly exercise every system that would remotely have a bearing on the success of the mission. Leo got the airplane up to about 93 or 94 knots on the last couple of taxi events.

Q: The C-5 is such a large airplane; why didn't you have more crew, at least one loadmaster in the rear upper deck as a roving patrol?

MITT: Leo and Lloyd Frisbee established many years ago that all airplanes must demonstrate airworthiness before the crew size can be anything but the minimum number required to fly the machine. This policy was set in stone and seconded by Shockley and Haley.

Q: Well, you had all the necessary parts; did everything work as advertised?

MITT: The basics did just fine. The engines and all the flight controls were great, however we didn't get the gear retracted because of some bad micro switches. The gear worked on the subsequent flights. We flew seven flights for a total of twenty (20) hours and one minute prior to declaring the airplane airworthy. However, it is also necessary to note that this first series of flight tests on all airplanes were flown without SAS or autopilot. The absence of these refinements did not materially alter the program. This was a prime example of maturing systems in a concurrent manner. That was the nature of the C-5 program as a whole. Neither the engine nor airframe were off the shelf items as you know.

Q: There must have been some unusual in-flight events.

MITT: We were airborne for one hour and thirty four minutes. In that short time Leo did the takeoff, flew for the first 45 minutes, turned it over to Walt Hensleigh who got the feel of things while Leo and Joe Schiele switched seats. After Joe was satisfied, I found myself in the seat and got to fly for a full twelve minutes; as a flight test engineer, this was not crucial to my education, but it puts events in better perspective for me in later tests (and conversations). Leo landed the airplane, and we proceeded to the post flight chaired by Tom May, who had quite a hiccup when I announced the takeoff time as 0747.

Q: Who were the crew on the other six flights?

MITT: Flight two had Walt in the left seat, Joe Schiele in the right and Hank Dees in the jump seat. Jerry and I stuck it out for the duration. The subsequent flight probably included Jesse Allen and Glen Gray from Lockheed; Jesse Jacobs, Colonel Ralph Matson and Joe Guthrie were alternating for the U.S.A.F.

Q: What kind of mementos were you carrying on the big occasion?

MITT: I'm certain everyone had something; Lockheed provided six model C-5As, which were made of aluminum. Each crewmember on the first flight has one, the sixth model went to Texas A&M.

Q: Did you have to stand in the relief crew compartment to run the instrumentation recorder all the time; didn't the chore get a little tedious after a while?

MITT: The recorders and associated equipment were located just inside the relief crew area, so there was a certain amount of movement into and out of the cockpit. I also was the roving observer. On one of my last flights at Marietta, Joe Schiele and I were patrolling the aft passenger compartments; it was on flight six or seven. I crawled into the un-pressurized section over the rear cargo doors, eased my way back to the vertical fin, and climbed up the ladder (which is permanently attached inside of the vertical fin, designed for ground maintenance use) to the top. A cold, noisy place, and I didn't tarry or put my initials up there. Just checking everything there was to see, a first and last for me.

Q: When did you leave to go to Edwards AFB?

MITT: I was GELAC's group leader, base commander, you could call it. Lockheed had as many as five C-5s at Edwards during the test cycle. Performance testing, systems verification, in-flight refueling tests and a whole variety of electronics tests. What made life interesting was the fact that one would run into a number of people who, through ignorance or mischief, would take verbal pot shots at the airplane. I had several long discussions with guys who felt that the IFR system was an overkill. When would an airplane that has a capacity of 325,000 pounds of fuel ever need to refuel in flight? That question has been answered many times on missions flown over the past thirty years. However, my response was that the weight trade-off between cargo and fuel is dictated by the variables of mission distance, field lengths, fuel availability, and that IFR capability covers a multitude of unknowns. Then questions would come up about the Ram Air Turbine (RAT). I guess they never knew about the four engine flameouts experienced in the early days of the DC-8 and other jets.

Jerry Edwards, Flight Engineer Senior

Q: What was your initiation into the C-5 flight program?

JERRY: I joined the project in early 1966. Like most of these assignments, the first contacts are with engineering to obtain detail knowledge on the airplane flight controls, engines, and other primary systems like hydraulics, electrical and bleed air systems. On a project this size there were numerous formal meetings that lasted from one to four hours. As the hardware became reality the pressure to get into heavy detail increased.

Q: You had flown quiet a bit on the C-141; the C-5 flight engineer's station is quite similar to that in the C-141 Starlifter, isn't it?

JERRY: The answer to that is yes and no. The big difference, of course, is the flight engineer's station is on the second floor, (with the rest of the crew). Seriously, the C-5 flight engineer's job expanded. His flight station layout is the best indicator of the number of changes. This airplane has two APUs, two ATMs; a more complex bleed air system to pressurize, air condition and deice the airplane using two pressurization/ air conditioning packs; there are dual fuel pumps in each of the twelve fuel tanks. Fuel usage has a strict progression for emptying the tanks. Then too, the FE has to manage inter tank fuel transfer. Unique to the C-5 is the hydraulic power transfer unit(PTU) which is used to pressurize any of the four hydraulic systems without transferring fluid, more controls to monitor. Also, the FE panel provides all the electrical systems controls. But the system that takes time to understand is the Malfunction and Detection Analysis System (MADAR). This gem is actually a very useful device that tracks specific systems and the engines. Data is displayed on a monitor which can be read by the operator. MADAR also has a memory storage device that will provide essential maintenance information in printed or schematic format. So, the answer to your question is that the new equipment and panel changes outweighed the small number of similarities.

Q: Didn't Lockheed set up a number of training classes to cover systems?

JERRY: Yes, eventually. But recall that this contract was production concurrent with development. This kept a number of airplane systems in a state of flux for a much longer period. In order to keep up, the flight crews did a lot of liason between offices in the B-1, the manufacturing location; the newly constructed B-95 design offices; the B-4 our old engineering flight test office; and then the new L-10 facility which was built on the south side of the Dobbins runway on company property.

Q: Since you were designated as a first flight crewmember about eighteen months before the 30 June first flight in 1968, what change in personal schedule did this cause?

JERRY: This was pretty exciting, considering the rest of the talent Leo could have chosen; my biggest change was working those twelve hour shifts. As the airplane came together, a flight test engineer, Louie Lamb, and I decided to work this way to assure 6001, the first airplane, round-the-clock flight crew coverage.

There is only so much one can squeeze from paperwork, and, although we never bucked the first rivet, we did help as all the systems were made functional. For example, this was my first opportunity to mess around with a landing gear system that castored, rotated and retracted. Naturally, we all spent a great deal of time developing the several methods to assure proper extension. It was difficult to get the gear to cycle properly in the early months prior to first flight. The original gear retraction system had numerous interface problems; thus Bob Altman and Henry Quay had their engineers design large piston type actuators to be fitted on the landing gear of the first two aircraft which assured proper rotation, alignment and retraction. The final design changes were tested and installed on a short term retrofit basis later.

Q: As first flight date approached, especially in the months the airplane is under flight operations responsibility, what systems get the most attention?

JERRY: You never felt that you could learn too much about the engines. Lockheed had its own engine test stand functioning at the flight line near the taxiway from the B-25 hanger; so we spent a fair amount of time at this test site. And the flight control system had 24 hour coverage for three weeks prior to that 30 June date. Louie Lamb, Whitey Kleven and myself shared the coverage requirement. Engineering and production people had to make the flight spoilers, ailerons and ground spoilers function properly in sequence. Remember that flap extension limits flight spoiler travel, and ground spoilers are mechanically locked out until the gear struts are compressed. It just takes time to get all these critical flight controls synchronized.

Q: The week before the first flight, the airplane taxied several consecutive days, why?

JERRY: As I recall, the pilots chose to get a feel of the flight controls under progressively higher taxi speeds. That activity also reveals idiosyncorsies of the brakes and anti-skid systems. Some fixes were accomplished while we waited on the airplane. Then too, we were looking in detail at electrical, hydraulics and bleed air systems under dynamic conditions for the first time. A host of inconsistencies were noted, fixed, or handed over to engineering for resolution, a fairly time consuming process. We lived in our flight gear for more than a week.

Q: Did you do any extensive engine run-up prior to the first take-off, anything similar to the two minute drill that came along later in the flight testing cycle?

JERRY: No, not this time. Leo set the throttles to the computed EPR and we rolled.

Q: Was the first flight exciting to you personally? After all, first flights set records, and, in this case, the world's biggest airplane set a flight mark that would stand for a while.

JERRY: In retrospect, the flight was a relatively easy task. It did not produce any serious problems; in fact, since the landing gear didn't retract on the first attempt, Leo said we weren't going to make any more attempts, and it remained extended for the full flight. We did take several engine power readings; we maneuvered over the Lake Lanier flight test area. This gave the north Georgia residents a first look at us; I don't believe we climbed any higher than 12,000 feet.

Q: There were dozens of news media people all over the Lockheed reservation, and there were family members here too. Did the holiday atmosphere last for more than one day?

JERRY: My family made it into the L-10 building. My children still talk about the event to this day. There were only five crewmembers on the first flight. Each had the opportunity to invite someone. In the post flight discussion, however, only the media were there, in addition to management and a dozen engineers. Leo, Walt and Joe handled all the questions. I personally think that "flight engineer" is a foreign language to most of the media. At any rate, it was a Sunday morning flight, and we actually got to leave the plant before dark that day. The serious side of business set in early on Monday, the first of July.

MITT: It should be mentioned that Lee Poore was on hand then. He arranged for a fine party at Squires Inn that evening. Since it took us a week of taxiing to get our nerve up he had to get the host location to be pretty flexible, only five reschedules, I was told.

Q: Did you do much flying at Edwards AFB during the first year of flight tests?

JERRY: None. My job was one of instructing and checking out the other flight engineers. The majority of my flying was accomplished from Dobbins AFB. Through a pleasant quirk in the teaching process, I flew on the first ten C-5s that rolled off the production line.

Q: Who were the next few Lockheed crew members transitioning to the C-5?

JERRY: In the pilot group Walt flew flights two and three. After that Hank Dees, Glenn Gray, Jesse Allen, Army Armitage and Carl Hughes all got opportunities. FEs were Malcolm Davis, Bobby Killcreas, Art Graves, Jack Parker and Jeff Fowler. Since everyone was growing up with the airplane the formalities of checkout were limited.

Editorial note. The ground and flight testing that was a concurrent process with the production of the airplanes resulted in numerous changes that were not easily incorporated in the B-1 production line. Lockheed established production modification operations at Palmdale, California.

Q: When did you go to Palmdale to flight test the airplanes in the modification process?

JERRY: In early 1970 I was there to prep the airplanes for flight plus do the test flights. I was on Ship 0011 when the fire occurred. That airplane never flew again; the fire damage was that severe.

Q: Can you give a few details?

JERRY: This is a combination of what I recall, combined with intelligence learned after the fact. We were taxiing on the runway, downwind, toward the takeoff position. This allowed several system checks. Homer Blaylock, the pilot, determined the rudder pedal nose wheel steering was inoperative and called the tower for clearance back to the flight line. We had a split crew, one Lockheed pilot and flight engineer and one USAF pilot and engineer on board. I turned the FE panel over to the USAF engineer and went into the cargo compartment to prep for parking. Smoke was beginning to fill the compartment. I immediately alerted Homer; he taxied clear of the runway while calling for fire trucks, set the parking brakes, shutdown the engines and cleared the airplane. Then we tried to help the firemen. The ATM, which has titanium fittings, was the fire source. Hydraulic lines were burned through, releasing gallons of fluid from reservoirs in the cargo compartment several feet above the ATM location. The spread of the fire eventually set the oxygen converter blazing also. It is also located in the wheel well. The intensity of the fire increased, severely damaging the fuselage rings, primary fuselage structure. Thus, the airplane was judged un-repairable and never flown again. The wings, engines, empennage, aft cargo doors, pressure door, visor, forward ramp, and cockpit instruments were all salvaged, as were, many smaller valuable components. It is believed that the wings were delivered to Marietta and put on ship 0081, the last C-5A.

Q: This was a major catastrophe; were there any personnel injuries?

JERRY: No, thank God.

Q: You flew on numerous C-5 flight test programs, which one stands out in your mind?

JERRY: The most memorable effort, and probably the most productive, was the flight series involving preparation and airdropping the Minuteman missile. The pressure to complete the missle launch was focused from well up the management ladder. After all, Secretary of State Henry Kissinger was waiting on successful results. That program really went like clockwork, but we had in-house skeptics. I personally know of one who had to eat crow. How about that Mitt?

MITT: Well, now that you mention it I do recall something. When the program was first proposed it came with so many caveats that I couldn't get excited in the least. Unfortunately for me, I expressed the "it will never happen" words to Chet Payne. And he wouldn't let me forget that momentary slip of the tongue. Months later, he made an unannounced presentation of the "bird" at dinner in El Centro before the last test airdrop. The whole crew just happened to be eating in the same restaurant that night. It's a good thing that I don't embarrass easily. By the way, I still have that piece of artwork on my mantle.

Q: Was that the only fun the Lockheed test director had on the Minuteman airdrop program?

MITT: Not the only one. Frank let me ride shotgun on a taxi maneuver at Ogden AFB. The Boeing troops were in a fit to begin loading the missile. The position of the airplane on the ramp had to be adjusted laterally about 40 feet. The easiest method was to move the airplane forward, taxi a circular route down a parallel runway and move into the new location. The taxiing actually took about ten minutes. Since other crew members were not showing till later in the day. I subbed as copilot.

Q: Well that doesn't seem like such a big deal.

MITT: Not to you perhaps, but due to a bungled preflight inspection by a phantom flight engineer we had two unauthorized passengers aboard, in the right wheel well. Fortunately, they had the good sense to wait till we completed our taxiing before deplaning. After the shock we all laughed a little.

Q: Didn't you get to the Paris air-show in 1972?¸

JERRY: Yes, The Paris Air show flights demonstrated the C-5 short field capabilities; Gray and Hadden had a ball showing how well the C-5 could perform. I never saw a more consistent series of short field touchdowns and short landing rollouts, one a day for six days; he hit the same tire mark on each landing six days in a row. The airplane did good, but Glenn was playing it like a virtuoso.

First liftoff, 30 June 1968. Lockheed photo.

U.S. AIR FORCE

Lockheed flight line.

CHAPTER 6

BIG AIRPLANE – BIGGER CHALLENGE

THE C-5A AND ITS NEW TECHNOLOGY

"Wide body airplanes appear massively cumbersome, probably lethargic. The C-5, the largest of them all, is agile, responsive—it is a flying Maserati."

E.B. (Gibby) Gibson

The C-5A airplane is a military transport with the capability of airlifting large, heavy, prime warfare equipment and the combat troops simultaneously. The fuselage design is a double deck structure. The cargo compartment (lower deck) will accommodate main battle tanks, armored personnel carriers, howitzers, helicopters, or small airplanes. The upper deck is divided into two compartments. The area forward of the center wing box is the flight deck which adjoins a compartment with fifteen airline seats, six bunks, a galley and toilet facilities for the flight crew and courier personnel. The upper deck rear compartment has 75 airline seats, with kitchen and toilet facilities.

To more easily grasp the size of the cargo compartment floor area, consider this. One C-5A can load six Greyhound buses, three pairs of buses in trail, with room to spare. They can be driven in the rear door, parked, and, when delivered, drive forward out the nose of the airplane.

The C-5 airplane is a high-wing monoplane with a 25 degree swept wing and a "T" tail empennage. The Galaxy is powered by four General Electric turbo-fan power plants, which are pylon mounted under the wing. Each engine is equipped with a cascade design thrust reversing system. All four thrust reversers can be utilized on landing after touchdown to reduce rollout distance. In flight, only the inboard reversers are utilized to achieve rapid deceleration and greater descent rates. Using reverse thrust, the airplane can be taxied rearward (backed) at gross weights as high as 730,000 pounds.

For comparison, and to fully appreciate the remarkable changes in airlift capability the C-5A represents, consider this quote made in December 1969 by Dr. R. L. Seamans, Jr., Secretary of the US Air Force, when the first C-5A became operational.

"The ton-mile operating cost of the C-5 is appreciably less than the older C-124 and C-133 propeller airplanes. It is even lower than the efficient C-141 jet transport and compares favorably with similar, large size civilian aircraft. In millions of ton-miles per year for each type of aircraft, one C-5A squadron is the equivalent of twenty one and a half (21.5) squadrons of C-124s, seven and one half (7.5) squadrons of C-133s, or four squadrons (4.0) of C-141s. This is an appreciable increase in efficiency and capability."

The C-141A, the first swept wing, turbo jet-powered, military cargo air-lifter, was delivered in the early 1960s to the USAF by Lockheed.

To emphasize the quantum leap in aircraft efficiency provided by the C-5A, a direct comparison of specifications on C-5A and C-141A is provided at right.

COMPARATIVE DIMENSIONS AND SPECIFICATIONS

	C-5A *Galaxy*	C-141A Starlifter
Length	247.8 feet	145.0 feet
T-tail height	65.1 feet	39.3 feet.
Wing span	222.8 feet	165.0 feet
Design Weight	764,000 pounds. (1)	316,000 pounds
	840,000 pounds (2)	
	920,000 pounds (3)	
Max Payload	265,000 pounds (4)	70,847 pounds
Engine model	TF-39-1A	TF-33-P-7
Manufacturer	General Electric	Pratt & Whitney
Thrust/engine	39,850 pounds (initially)	21,000 pounds
	43,000 pounds (engine upgrade)	

Notes

(1) The gross weight (GW) limit at a load factor of 2.25g
(2) The GW limit with the new wing installed; Also the C-5B capability.
(3) The GW flight limit with in-flight refueling. C-5A & B capability
(4) The maximum payload for the C-5B and re-winged C-5As.

Cargo Loading Features

The C-5 has forward and aft cargo doors, which permit drive-through loading and unloading. The nose door has a visor-like appearance when in the open position. The forward door complex includes permanently installed ramps. These ramps, when extended to the ground, have an eleven-degree incline for drive-in loading. Ramps can also be positioned to truck-bed height. In the fully opened position, the visor nose clears the cockpit windscreen allowing pilot visibility for taxiing, if desired. The aft center door can be opened in flight for airdrop missions; the airdrop speed limit is 205 knots (235mph). Two full access side doors, for ground use, operate in combination with the center door for full width vehicle, container, or palletized cargo loading. The aft center door, which is 19 by 13.4 feet, can be extended to the ground in a similar manner to the forward ramp. These ramps support the loading of equipment including the 128,240 pound bridge launcher.

Cargo loading is simplified by the C-5A's capability to lower the cargo floor to 79 inches above the ground from its taxi height of 113 inches. This unique "kneeling system" is hydraulically powered. The system can be pressurized by the engine driven pumps, or pumps driven by the auxiliary power units (APU). Each main landing gear pod houses one APU.

Aerial Delivery

The C-5 aerial delivery capability allows sequential extraction of palletized cargo for the airdrop mission. The individual pallets with cargo can weigh up to 50,000 pounds each. Paratroop doors are located aft on each side of the cargo compartment, forward of the aft cargo ramp. Jump platforms and air-deflector doors are utilized during paratroop missions.

Fuel System

The fuel system consists of twelve integral fuel tanks, six in each wing, designated as main, auxiliary, and extended range tanks. The maximum capacity of the fuel system is 51,150 gallons. When a full fuel load is aboard, each wing tip will deflect downward 1.5 feet from its static, empty tank, position. The system is designed so that each engine is normally supplied fuel from its related main, auxiliary or extended range tanks. Capability exists to transfer fuel from any auxiliary or extended range tank to any other main tank or engine. The ground-refueling rate is 600 gallons per minute. Two single point refueling receptacles (SPR) are located in each wheel well. The C-5 is in-flight refueling (IFR) capable. The receiving coupling for IFR is installed in the upper fuselage over the cockpit. Tanker airplanes are the KC-135 or KC-10, equipped with controllable booms for the IFR mission.

Flight Controls

Primary flight controls are the elevators, rudders and ailerons. Five dual-purpose flight spoiler panels on each wing are mechanically synchronized to the ailerons to achieve the desired roll response, with spoiler gearing and up-rig increasing with flap deflection. For directional and pitch control, each primary control is divided into two surfaces (two rudders, four elevators) with each "half surface" controlled by different mechanical linkages and powered by different hydraulic systems. This provides a jam-proof mechanical system and fully redundant powered controls. Directional and lateral trimming are accomplished by changing the null position of the rudder or ailerons. Pitch trim is accomplished by angular positioning of the single structure horizontal stabilizer.

The secondary controls consist of leading edge slats, Fowler flaps, ground spoilers, and a stall-limiter device. The stall-limiter's purpose is to alert pilots when they are approaching a critical slow airspeed regime.

Electrical Power

Four engine driven AC alternators provide 60/80 KVA power. This primary power source is backed by two APU driven generators of equal capacity. DC power is available from the secondary system. Two 5 ampere hour 24 volt batteries are also installed. For emergencies, the air turbine motor (ATM) located in the left wheel well can be extended in-flight to provide hydraulic pressure, and AC and DC electrical power.

Environmental Control System

The Galaxy environmental systems provide air-conditioning and pressurization. A cabin pressure altitude of 8,000 feet can be maintained to 40,000 feet flight altitude. The flight crew, cargo, and passenger compartments are pressurized simultaneously.

Landing Gear System

This landing gear system has several distinctly useful qualities beyond its routine purpose. Four main gear bogies, each carrying six wheels in a tricycle pattern, and a nose gear with four wheels abreast, provide ground maneuvering capability over both paved and unpaved ramps, taxiways and airports. This high floatation design allows ground maneuvering on fields with a compaction measurement as soft as 9 on the California Bearing Ratio Scale (CBR-9). The twenty-eight wheels and tires are of identical size. The nose wheels can be turned 60 degrees either side of center by a manually operated steering wheel located on the pilot's side. Rudder pedal nose wheel steering is also available; this limit is (+/-) five degrees from center. The aft main bogies can be castored (hydraulically positioned to either left or right of the taxi direction); this reduces the taxiing turn radius, allowing the C-5 to reverse the direction of taxi in less than 150 feet of runway or taxiway width.

The kneeling capability of the airplane lowers the cargo floor in a forward only, aft only or simultaneous (level floor) method. Independent bogie lifting (reverse kneeling) facilitates landing gear and wheel brake maintenance. Under reduced gross weight conditions, the C-5 can take-off and land with one or two bogies retracted. A tire pressure deflation system is operable from the flight deck to reduce tire pressures when landing is planned on unpaved runways. Crosswind (CWS) gear capability was provided on the first eighty-one C-5s, but operational experience proved this system to be unnecessary on the C-5B aircraft.

Hydraulic Systems

Four independent hydraulic systems, each powered by two engine driven pumps provide the muscle to operate the primary and secondary flight controls, the wheels brakes, the fore and aft cargo doors, and the kneeling system. This hydraulic system can deliver pressure without fluid transfer, by means of interconnect motors; thus, one hydraulic system can pressurize two or more systems. Each APU, one in the left wheel well, the other in the right, powers a hydraulic pump to pressurize its respective number one or four hydraulic system. (1)

Typical Cargo Loads

Typical non-lethal and combat cargo loads are compared below

Non-Lethal Cargo	
One D-8 tractor	41,820 pounds
M-35 truck & tractor	23,175 pounds
3 two ton fork lifts	19,800 pounds
Three Lark V's	60,000 pounds
Refrigerator truck	41,913 pounds
Troop kit equipment	9,000 pounds
M-43 ambulance/trailer	12,380 pounds
TOTAL WEIGHT	219,768 pounds

Combat Cargo	
Two MA-1A tanks	each 110,000 pounds
Six crew members	1200 pounds
– or –	221,200 pounds
Eight armored personnel carriers	
and crews @	11,760 pounds
	94,080 pounds

Cargo loading is a critical function for success in airlift, regardless of cargo weight or size. Unit weights vary from a 6,300-pound forklift to the 128,240-pound bridge launcher. Loading and unloading routines are critical to mission completion in acceptable times. The flexibility of the C-5 to rapidly complete loading or unloading is illustrated by the following five figures. Figure 1 shows a

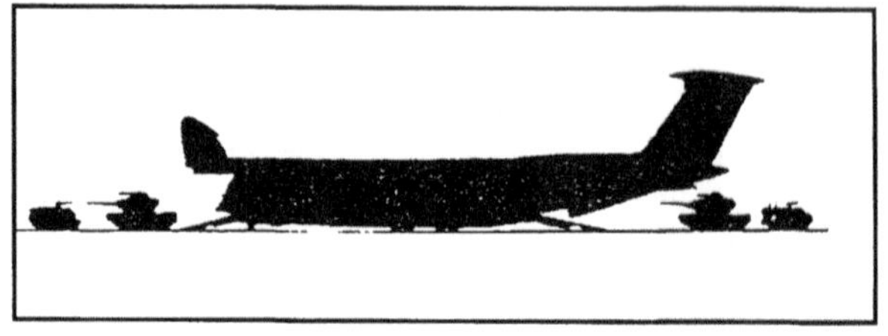

Figure 1

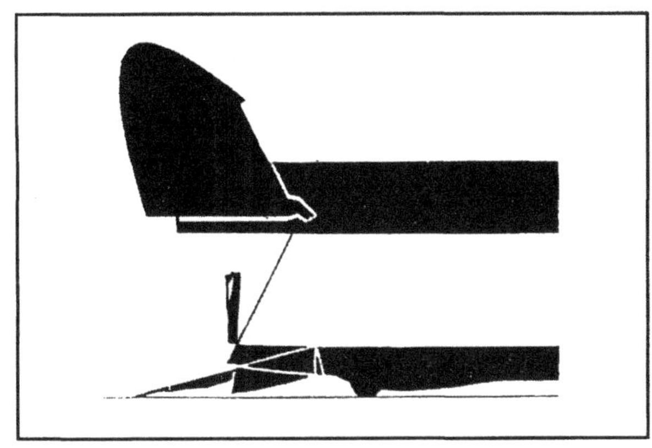

Figure 2

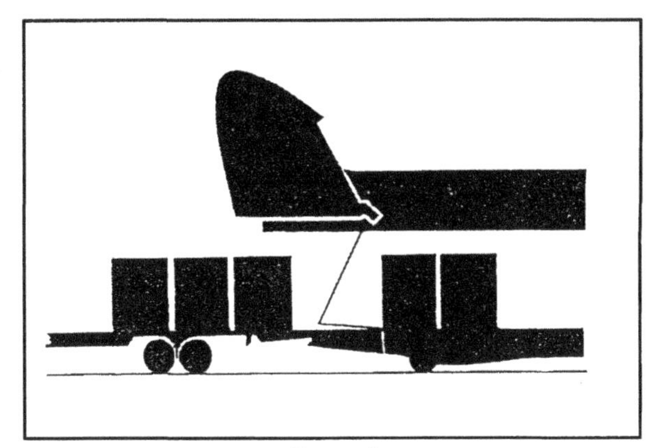

Figure 3

Figure 4

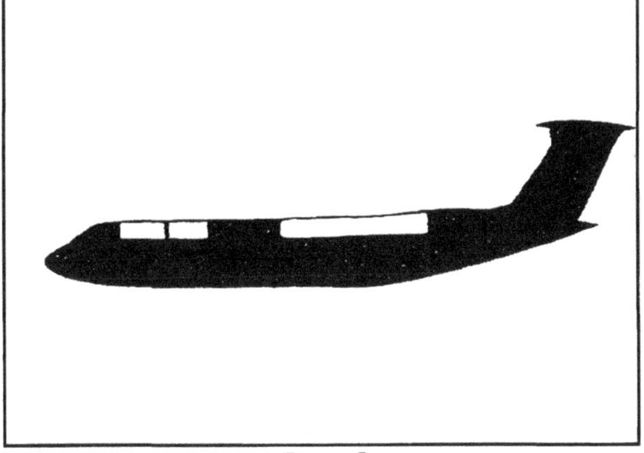

Figure 5

profile of the airplane with the forward and aft doors open, permitting drive-in and drive-out capability. During the May 1972 North Vietnamese offensive, C-5s delivered both M-41 and M-48 tanks to Da-Nang AB in South Vietnam, completing touchdown-to-takeoff times of less than forty minutes. In addition to delivering combat equipment in ready-for-immediate-use status, rapid off-loading also greatly reduced the risk of enemy ground action against the airplanes.

Figure 2 shows the C-5 in the kneeled position, with its full width, self-deploying, integral cargo ramp extended to the ground. Figure 3 depicts the truck-bed configuration of the ramp loading palletized cargo. It is important to emphasize that the C-5 is self sufficient in these critical evolutions. Ground equipment is not required to position the airplane or to configure it for these loading or unloading tasks. Figure 4 illustrates the side by side loading method used for airlifting multiple wheeled cargo or pallets. The US Army prefers to carry divisional bulk cargo to whatever extent practicable on their vehicles being airlifted. The C-5, because of its cargo compartment size, 19 feet wide, 13.5 feet high, and 121 feet long, can mix classes of cargo. A variety of equipment remains intact on one carrier. Figure 5 shows troop and crew seating locations. The U.S. Army found when vehicle crews accompany their equipment the probability of mission success is significantly raised. The available seating, fore and aft, exclusive of basic aircrew flight deck seats, is 90 airline style seats. This capacity is available without utilizing necessary cargo space.

Engines

The General Electric TF-39-1A engine, which powers the C-5, is a turbo-fan, dual rotor, variable stator high-bypass ratio, and front fan engine. The one and a half stage, front mounted fan, produces 85% of the engine thrust. It is driven by a six-stage, low speed, turbine. The core engine has a 16-stage compressor, an annular combustor, a two-stage turbine, a gearbox, controls and accessories. The original engine, rated at 39,850 pounds of thrust, has been upgraded to 43,000 pounds for the current C-5 fleet. This eight to one bypass ratio engine weighs 7,884 pounds dry weight. At maximum power the airflow is 1,550 pounds per second.

The engine is 16 feet long; the pod length is slightly over 26 feet long with an outside diameter of eight and one half feet. The engine specific fuel consumption (SFC) at takeoff thrust is 0.327 lb./lb./hour. During optimum cruise, at Mach number of 0.767M, the SFC is 0.582 lb./lb./hr. (2)

New Technology Incorporated in the C-5A

The C-5 airplane represented significant advances in technology as it entered service in December of 1969.

Aerodynamics

- The airfoil design achieved a higher drag divergence Mach number for a given wing thickness ratio; the Hoerner wing tips reduced wing tip vortex drag at all speeds.
- The wing fillet and MLG fairing locations minimized aft fuselage flow separation.
- The engine nacelle location and inclination achieved an equivalent cruise drag reduction of five percent.
- The leading edge and trailing edge flap system produced the highest lift coefficient for a swept wing ever achieved to date. The flap track fairings actually reduce the airplane cruise drag because of a newly discovered, and now patented, aerodynamic principle.
- The fastest roll response of any jet transport aircraft, regardless of size, to date.

Structures

- The fabrication of mainframe forging introduced a product size of 6,000 square inches, double those previously offered to industry.
- Chemical milling process, a new technique to effect close tolerance shaving of excess metal from fabricated products.
- Primary and secondary structures of metal bonded materials.

Functional Systems

- Torque limiters, which are load sensitive brakes used in mechanical systems, prevent permanent damage due to overload.

Fail Safe/Multiple Controls

- Multiple aerodynamic control surfaces are installed on each axis. Each flight control system featured dual, independent, hydraulic power sources, parallel mechanical load paths, and mechanical jam protection.

Automatic Flight Controls

- The autopilot system provided automatic Category II instrument approach, landing and rollout capability. Several incremental modes in the auto pilot system were added.
- The Stability Augmentation System dramatically improved flying quality characteristics to equal, or better, those of much smaller, lighter aircraft and allowed a smaller horizontal stabilizer for lower empty weight and reduced drag, without sacrificing handling qualities or flight safety.

Electrical Power

- The electrical power generation system produces 1 KVA per 1.9 lb. of system weight, compared to 4.0 pounds per 1 KVA on other aircraft.

Hydraulics

- The power transfer units transmit hydraulic pressure between the four systems without transfer of fluid. The systems are reversible.

Landing Gear, Wheel Brakes

- Six wheels on each of four bogies provided high floatation on any landing surface and a universal in the bogie centroid insures all six tires touch the terrain with inflated or flat tires.
- The kneeling system allows the cargo floor to be lowered to truck-bed height.
- A single main landing gear can be raised off the ground to facilitate maintenance.
- Main landing gear shock struts are dual chambered.
- The first use of solid beryllium brake discs occurred on the C-5A.

C-5A Flight Characteristics

Today flying occurs with such frequency, the activity has taken on a "so what" character in minds of the public and crewmen alike. However, this casual note does not diminish important realities. Flight involves profound transition. The moment he becomes airborne, the pilot leaves a forgiving arena for a zone where time is dominant. Fuel remaining, unseen traffic, weather extremes, advancing miles in seconds. These factors dominate thought. Precision is a goal, focus an imperative, one plans for no second chances.

There are several reports in trade magazines, such as Aviation Week, Air International, the USAF's own in house publication, "The MAC Flyer" and a Shell Oil company publication describe the C-5's flight characteristics in short, generalized terms, somewhat resembling "sound bite" technology in print.

"Great, easy to fly."

"A marvel, considering its size."

"Stable, yet responsive; a great instrument airplane."

"Designed to fly easily into short airfields; it fulfills that promise."

The balance of the story tends to dwell on size and all those related concerns that "being the biggest" must address. Then, there was a story in Air and Space, September 1989, that had so many fundamental mistakes the USAF felt obligated to instruct the author in writing (5).

Simplistic flight descriptions are generally satisfactory for the average situation; however the absence of details blocks real understanding of how well the Galaxy can fly. The next few paragraphs should help fill that void.*

There were numerous requirements in the contract defining how the C-5 must perform its variety of missions. A prime governing document was Military Standard 8785 (MIL-F-8785), which is the airplane flying qualities regulation. In the early design stages for the C-5, it was found that although the demands of MIL-F-8785 were apparently met, fixed based flight simulator results as reported by company test pilots gave the airplane poor Cooper-Harper ratings in a number of configurations. Thus, simulator guidance took on a higher degree of importance as the C-5 airplane configuration was optimized.

In the early 1950s two NACA research engineers developed a method for rating airplane-flying qualities. Cooper and Harper labored long and hard to quantify what is inherently qualitative evaluation. "Flies great" or "A real truck", or similar descriptions are not capable of the level of accuracy or detail required in evaluating increasingly complex aircraft. The flexibility and communicative accuracy presented by this numerical coding quickly gained universal acceptance. The language of the Cooper-Harper (CH) rating system for airplane flying qualities is very precise. The CH system assigns task oriented flight segments numerical ratings from 1 to 10. Each number has attached to it a precisely worded description that leaves almost nothing to the imagination of the reader. CH-1 rating is the best possible grade. CH-2 and CH-3 indicate progressive flying quality degradation. CH-4 to CH-6 points out deficiencies of increasing seriousness. CH-7, 8 and 9 ratings identify safety issues requiring immediate attention. The CH-10 rating indicated a totally unacceptable quality, necessitating immediate grounding. In the final operational configuration the C-5A earned a CH-1.5 rating. Certain mission configurations received slightly less favorable ratings when the flight gross weight exceeded 728,000 pounds.

Two flight conditions that normally tell the tale of an airplane's maneuvering capability, though not too easily understood, are: (a) the airplane flying qualities in the landing configuration and, (b) the flight characteristics near or in the stall region of the flight envelope.

---

* *Notable comprehensive writeups of the C-5 by then-Major Don Fremming and Captain D.W. Traynor were published in USAF publications.*

Landing Configuration Flight Characteristics

A prime flying qualities requirement for the C-5A was eventually verbalized and became known as the "sidestep maneuver". It was defined in this form.

"In any reasonable set of weather conditions, the crew must land routinely after breaking out of the overcast at 200 feet above ground level, and finding the airplane displaced 200 feet laterally from the centerline of the runway".

This set of conditions poses a very rigorous challenge, considering that the wingspan is 222 feet, (that is, the wing span is greater than the ground clearance) and the gross weight will be generally be well in excess of a half million pounds. However, the landing can be completed routinely, if the correct controls are applied promptly. Applying only 30 degrees of control wheel rotation, banking to 13 degrees at a roll rate of four(4) degrees per second will maneuver the airplane to the runway centerline. Applying opposite controls to stabilize on centerline will achieve the desired airplane positioning. Simultaneous application of the needed pitch control to flare is easily done in this maneuver. The control harmony in the lateral and pitch axis, a measure of the airplane response and necessary control force required to maneuver, is rated superior in the C-5A.

This landing configuration maneuvering requirement grew out of a contract line item which stated the C-5 must achieve a bank angle of eight degrees in the first second following a full roll command. It is clear that, even though the airplane was designed for cargo missions which are stereotyped as "straight and level for hours at a time", the US Air Force intended the biggest airplane in its inventory to be agile, responsive and capable of maneuvering in a compatible manner with other smaller aircraft around any airport facility.

An example of the excellent landing capability might best be described by it's performance at the Paris Air Show at the Le Bourget airport in June 1971. At a gross weight of a half million pounds the airplane made takeoffs using 1700 feet of ground roll, rotated at 93 knots, lifted off at 100 knots, and climbed to pattern altitude at 111 knots. After completing a high speed fly-by at 300 knots, then, a slow speed pass at 110 knots, the short field landing capabilities were demonstrated. After touchdown, three hundred feet beyond the approach end, a full stop landing was completed requiring only an additional 1500 feet of runway (thus, the ground roll was only six times the length of the C-5). The rapid deceleration on the ground is possible by proper use of the ground spoilers and maximum anti-skid braking. Engines were positioned only to idle reverse power during the rollout. The exceptional effectiveness of the C-5 anti-skid brakes minimizes the use of full reverse thrust on landing, especially when the airplane gross weight is low and runway length is not marginal. (The C-5A 1-1 illustrated that wheel brake effectiveness contributes more than 90% to the rollout performance.) The company pilots who flew these demonstrations were Glenn Gray and Frank Hadden. (ref. A.W. 28 June 1971) The demonstration airplane was leased from the 437th MAC wing based in Germany. Thus pre-air show practice time was zero. Though this short field capability was demonstrated very early in the life of the C-5 Galaxy, its utilization in military operation life has been minimal.

Flight at High Angle of Attack

The airplane does not exhibit any undesirable characteristics in the high angle of attack, slow speed area of the flight envelope. Wing roll-off is not encountered, little if any buffet is noted if the stall entry rate is at one knot per second deceleration. Designers tailored the C-5A to have neutral stability for about six degrees of the stall regime. If angle of attack is allowed to increase, the stability characteristic becomes mildly restoring. There is no tendency for loss of pitch control in any configuration, with the flaps retracted or extended. The C-5 does have a stall warning system, which includes a stick shaker.

The flight control system, which makes flying the Galaxy relatively easy, is described in Chapter Six. However, the complexity of the roll control system that results in such exacting flight performance is worth reviewing. The conventional ailerons are paired with five flight spoiler panels per wing. The flight spoiler deflection angle limit changes from a 22 degree clean wing limit to 60 degrees when the flaps are extended for approach or landing. Also, these ten flight spoilers open to a two-degree up-rig position when the flaps/ slats extend. Then, to achieve a desired roll rate, the left wing spoilers in a left roll (turn) will open to the commanded position; while, the spoilers on the right wing close completely. This spoiler action reduces the lift loss in effectively allowing the net balance to roll the airplane at the desired rate. The pilot's control wheel commands full roll authority with sixty degrees roll movement. This useful change from a conventional ninety degrees of wheel motion, as in the C-141 and 120 degrees in earlier airplanes, has a positive effect in reducing pilot workload. Pilot force required to move the control wheel is generally below 20 lbs. This important characteristic further enhances the good flying qualities.

For pilots transitioning to the C-5 from other airplanes having the cockpit literally on the second floor, 35 feet above the ground, has proved not to be a big hurdle. The excellent outside visibility, the simple engine controls, and responsive wheel braking system, enhance the "quick study" most crews exhibit in transitioning. After a half dozen flights the cockpit environment is quite normal. The C-5 cockpit is engineered to have essential flight data within easy scan range. The instrument panel flight information was arranged using conventional methodology. For the 1960's time frame, the C-5's use of a tape presentation of the flight and engine data was new and novel, but effective. Controls for the operation of landing gear, flaps, ground spoilers, and autopilot are well within the reach of either pilot. Cockpit windows are large providing both pilots excellent visibility while maneuvering on the ground. For emergency egress purposes, each pilot has a sliding window near his outboard shoulder. An escape reel, rigged with a steel cable is stowed below the window aiding in emergency egress from the cockpit.

In the C-5, the pilot is approximately 65 feet above the runway at touchdown with the airplane in the flared landing attitude. The completion of the landing and rollout are tasks requiring normal controls operated in the traditional fashion. Thrust reversing, when utilized, is controlled through a throttle "lift over a one inch gate" move; then pull the throttles rearward to obtain increased levels of reverse thrust. The lag between of throttle command and engine response is very small. Full power response is achieved in six to nine seconds after a rapid throttle motion from idle to the maximum power position.

Senator Barry Goldwater, using Lt. Col. Jesse Jacobs' headset, flies the C-5. Lockheed photo.

Loading a Chinook Heilo onto a C-5 at Cam Rahn Bay.

Chapter 7

Timely Airlift Meets Our Nation's Objectives

C-5s Perform Dominant Flight Tasks

These two short stories are part of the continuing history of flight operations of the C-5 airplane, emphasizing its usefulness in sustaining the nation's political and economic interests.

Operation Nickelgrass, October 1973

On 14 October 1973 the first of several C-5A's landed at Lod International airport in Tel Aviv, Israel, with a cargo of war material sorely needed by the Israelis to fight the combined attacks of the Syrian and Egyptian troops that invaded on 6 October. The invasion had occurred when the Jewish State was celebrating Yom Kippur, one of their most holy periods of the year. This action was the most serious political crisis in the Middle East since World War II. At least one commentator noted that the strength of the forces opposing Israel were equivalent in size to the NATO forces in Europe.(1) The possibility of attack was not completely unforeseen; however, Israeli intelligence sources had reported for several days before 6 October that both Egypt and Syria were sending troops to reinforce their borders with Israel. At first interpreting this as nothing more than military exercises, Israeli leadership took no action until hard evidence pointed to imminent attack. Reserve forces were called to duty a mere four hours prior to the first attack.

Beginning operations against Israel in the morning of 6 October, Egyptian forces crossed the Suez Canal at three points and moved into the Israeli occupied Sinai Peninsula—which Israel had captured from Egypt in 1967 in the Six-Day War. Northeast of this attack, the Syrians overran Israeli troops at two points in the Golan Heights, another territory Israel had taken in 1967. The Arab troops, estimated at 350,000, outnumbered Israeli forces four to one at the war's start; the Arab coalition had three times as many airplanes and tanks (2,800) as Israel. Not surprisingly, aggressor forces scored stunning victories in the first days of the war; one Arab high tech weapon was the wire-controlled TOW missile, extremely effective against the IDF armor.

Within days, Israel lost control of most of the east bank of the 103 mile-long Suez Canal, and Syrian forces had penetrated five miles beyond Israeli lines on the Golan Heights.

The outcome of the war swung in the balance for several days. Soviet-supplied missiles were effective in checking the Israeli air offensive against Arabic positions. Deciding against a two front war, during the first week of hostilities Israel fought a holding action against the Egyptians and concentrated on the Syrian positions. On 10 October 1973 the Israelis counterattacked with good success, repulsing Syrian forces in the Golan Heights and pushing eastward to within 30 miles of Damascus. With the Syrian front stabilized, Israel shifted forces to the Sinai, initiating a surprise assault on 17 October. By the end of the second week Israeli forces had broken the Egyptian lines in three places. The Israeli army eventually moved to within 50 miles of Cairo, nearly trapping the aggressor Third Army in the Sinai.

This conflict was a classic land battle of tanks and infantry, air superiority and interdiction. The two superpowers, the US and the Soviet Union, had opposing vested interests in the outcome of the conflict. Primarily, the Yom Kippur War was a war of attrition, as each side exhausted itself fighting the other. Within a week, combat materials were in short supply on both sides. This condition led directly to involvement of the US in the conflict as Israel's arsenal and the C-5 as the principal vehicle for early delivery of large heavy weapons to the scene of battle. As the war escalated beginning on 10 October, the Soviet Union airlifted war material to their clients, Egypt and Syria, using AN-12 and AN-22 transport planes. At its peak the Soviet airlift flew 70 aircraft a day into Egypt and Syria.(2). Given the Soviet airlift, many US political and military leaders believed American assistance to Israel was necessary to bring about a desirable resolution of hostilities. As diplomatic efforts failed to end the fighting, the United States agreed on 13 October to re-supply Israel with war material. Losses by the Israeli at this time were 4,000 dead or wounded, 110 aircraft destroyed, and more than 800 tanks lost.(3)

When this conflict began, the US military were beginning the readjustment to peacetime conditions following the full withdrawal of American forces out of Southeast Asia. This included a large reduction in personnel and resources of the armed services. Despite these circumstances, President R. Nixon wrote confidently in his memoirs that when news of the Israeli losses were given he stated "I had absolutely no doubt or hesitation about what we must do. I met with Kissinger and told him to let the Israelis know that we would replace all their losses, and asked him to work out the logistics for doing so" (4). This was in character for the Nixon administration, which publicly declared the US favored support to friendly countries by providing military equipment and supplies needed for self-defense whenever attacked.(5) The Arab world, as a whole, was intensely opposed to any assistance that might be provided Israel. Though unwilling to begin hostilities against the United States because of this crisis, they threatened (and carried out) economic sanctions, in the form of an oil embargo, because of the American actions. Representatives of large oil companies lobbied the President against assisting Israel, because it would lead to interruption of the flow of oil. Economic effects would increase the US and other western nations' dependence on Middle Eastern oil. Additionally, large insurance carriers refused to cover aircraft and ships that might become involved in an airlift that would enter a war zone. This prompted commercial cargo carriers to refuse to work in any re-supply effort. An added problem arose when several European nations in fear of Arab sanctions and the dicey diplomatic situation, refused the United States aircraft clearance to fly through their airspace with equipment bound for Israel(6).

Within hours of the decision to aid Israel, the Department of Defense (DOD) determined that the re-supply effort must be a two-pronged military movement of supplies to the stricken nation. The first part, and ultimately the most significant, was the airlift conducted by the Military Airlift Command (MAC) headquartered at Scott AFB, IL (7). This organization was born of necessity on 29 May 1941, as the Air Corps Ferrying Command with the duty to fly aircraft to the United Kingdom as part of President F.D. Roosevelt's lend-lease program. This organization graduated through several name and mission changes, emerging in the 1970's as MAC with a charter to accomplish all military airlift for the United States Air Force (8). The plan to use airlift was largely the result of time constraints; airlift could pickup and deliver equipment within an 18-hour window. Sealift, the second prong of the re-supply task, was used later; however, ocean vessels required 12 to 14 days en-route time. They were too slow to affect the outcome of the war. *Air transport was the only method of timely delivery.*

Enter the C-5A as the vital component of airlift to Israel. This airplane had been developed in the 1960's to transport rapid-reaction forces, primarily airborne groups, major battle equipment and heavy tanks to any theatre of the world. The C-5A has a payload of 240,000 pounds, twice the capacity of any other MAC aircraft in inventory at that time. The C-5A Galaxy is capable of delivering the immediate impact heavy armor, two main battle tanks, each weighing 105,000 pounds. With the C-5A, the U.S. Army would no longer need to give up a sizable portion of its firepower when relocating by air. Aerial refueling capability gives the C-5A unlimited range, enabling the projection of military forces to any portion of the globe. MAC received the first of its C-5A planes in December of 1969, and it proved itself a capable air-lifter during the Vietnam conflict of the early 1970s. Its strengths were again utilized as a principal air-lifter for fighters, tanks, and helicopters during the Yom Kippur War. (10)

National leaders approved plans for MAC aircraft to transport Israel-bound cargo to the east coast, mainly to Oceana Naval Air Station, for pickup by Israeli commercial aircraft. EL AL, the national airline of Israel, was able to move 5,500 tons to Tel Aviv using eight B-707 and B-747 aircraft, but this effort was abandoned since the most critical military cargo could not be transported in these aircraft. This situation ended American efforts to keep its military aircraft out of the conflict. President Nixon decided on the use of MAC aircraft when Israel asked the US for assistance in delivery of aircraft, armed personnel carriers, artillery ammunition, main battle tanks and comparable weaponry.(11) The large, heavy cargo could only be airlifted in the C-5A Galaxy. Thus, Headquarters MAC developed a plan to accomplish this movement of war material, which had the C-5A as principal air vehicle. General Paul K Carlton, the MAC commander, had his staff explore several options. The initial plan called for MAC aircraft to move approximately 4,000 tons of cargo(250 planeloads) to Lajes Air Base, in the Azores. The Israeli commercial planes would on-load the equipment and fly it to Israel. This would satisfy the national leadership's desire to maintain a low profile while providing aid. Questions remained concerning the availability of Lajes. The United States put pressure on the host nation, Portugal, to secure landing rights at the base for movement of this cargo bound for the Israelis. Without Lajes, the airlift volume would have been reduced to a trickle.(12)

As DOD considered various options Israeli Prime Minister Golda Mier sent an urgent personal request to President Nixon on 12 October asking the U.S. to deliver to her country the needed war material. Nixon instructed the Secretary of Defense to begin the airlift re-supplying Israel. Anticipating the pending transport directive, at 1950 Greenwich Mean Time (GMT) on 12 October, MAC activated its Contingency Support Staff at headquarters, placed all C-5 and C-141 air crews on Bravo Alert, established Support Staffs at the numbered Air Forces, canceled all but essential training, and designated the Twenty First Air Force as the MAC controlling airlift. On 13 October MAC had nine aircraft en-route to the east coast, complying with instructions to arrive and depart under the cover of darkness. The use of Lajes Air Base was still pending. On 13 October Defense Secretary Schlesinger directed MAC to fly the cargo all the way from the United States to Israel.(13) Briefed on the use of C-5s for the entire route to Israel, Nixon immediately agreed, commenting "Do it now". Kissinger recalls that at 12:30 PM in the 13th he told Israeli officials that the United States "would fly the C-5A airplanes direct to Tel Aviv until we had the charter (of commercial aircraft) issue sorted out, that supplies already in the Azores which were beyond EL AL's capacity would be moved to Tel Aviv in C-141s, and that Israel would shortly be receiving 14 F-4 Phantom fighters." (14)

The decisions at the highest levels of American government were acted upon by MAC personnel stationed worldwide. At 1645 GMT the same day, MAC received orders to fly the cargo to Lod International Airport, Israel, with no delay and irrespective of the cover of darkness. Fifteen other C-141s en-route to or already at east coast airfields were also ordered to depart immediately. By this time successful pressure on the Portuguese government had ensured that Lajes Air Base would be available as an en-route refueling stop. However, current 50 knot cross winds at Lajes caused a weather hold situation from 1645 until 2100 GMT on the 13th of October. The command developed a logistical pipeline reaching halfway around the world.

Equipment and supplies bound for Israel were gathered from 29 different locations in the United States, principally loading at military bases. Additional materials were withdrawn from the pre-positioned stores in Europe maintained by NATO. The initial flow rate for the airlift's first 24 hours into Lod was set at one C-141 every 40 minutes and one C-5 every 4 hours; thereafter, four C-5s and 12 C-141s were scheduled per day. The rate was adjusted upward to six C-5s and 17 C-141s during a 24 hour period on 21 October. Because of the distance involved, the need to pick up material at several locations, and the potential hostile environment, MAC used staged or augmented crews for the airlift. The C-141s used staged crews; they flew up to 18 hours between rest periods. The C-5 augmented crews, requiring a second engineer, navigator and loadmaster flew up to 24 hours prior to a rest period. Once loaded, the MAC transports completed the six-hour flight to Lajes Air Base in the Azores. The average flight length from a US base to Lajes was 3,300 miles. Lajes was the only site in Europe available for supporting this airlift, and, therefore, became the great staging base of this operation. This airlift constituted one of the heaviest levels of activity ever seen in its history, handling between thirty and forty missions per day for a thirty day period.(15) Mac deployed 1,200 additional personnel to support the operation. Major General W.E. Overacker, a C-141 pilot during this airlift stated, " there was plenty of spirited and imaginative improvising at Lajes to get the job done and we did the job indeed!"

From Lajes the MAC transports flew toward the straits of Gibraltar, then east over the center of the Mediterranean Sea, carefully avoiding territorial airspace. When reaching the Isle of Crete territorial limit, they turned southeast toward Israel and touched down at Lod International Airport, Tel Aviv. The requirement to avoid violation of foreign airspace complicated in-flight navigation. The US Air Force's command and control net, the air traffic control units of various nations, and the U.S. Navy's Sixth Fleet worked together to insure the MAC transports never deviated from approved flight track. The Navy stationed a ship every 300 miles along the corridor flown by MAC to backup routine navigational aids, and an aircraft carrier was stationed every 600 miles to respond in case of attack or harassment by anyone. Near Crete, the American and Soviet flight paths crossed. The US jet powered transports flew several thousand feet higher than the Soviet prop aircraft. Greece requested the United States to alter its flight tracks south of Crete to reduce the in flight collision potential. This was accomplished. In flight, communication between the Soviet's Egypt and Syria bound pilots and the American pilots headed for Tel Aviv was not uncommon. Fighters from the Israeli Defense Force (IDF) met the U.S. transports 150 miles from the airport to provide armed escort to touchdown. The Lajes/Tel Aviv leg averaged seven hours. (17)

Two airports, Lod International and El Arish in the Siani, were the offloading points for these airplanes. The automated cargo handling equipment arrived in the first planes, along with the staff for the Airlift Control Element (ALCE), which does scheduling, establishing the communication net, and coordinating the materials disposition. ALCE organized the Israeli work coverage to insure rapid turnaround for the MAC fleet. During the airlift 566 aircraft delivered 22,300 tons of material at Lod Airport. Armor, ammunition, weapons, planes, and helicopters were 21,190 tons of that total, the remainder being cargo-handling equipment. The airlift lasted a total of thirty-two days, 14 0ctober to 14 November.

This list detail the tonnage airlifted by both the C-5A and the C-141.

	C-5	C-141
Cargo weight: TONS	10,673	11,632
Number of flights	145	421
Average tons per mission	72.86	27.85
Number of pallets	2,644	3,564
Total flight hours	4,816	12,958
Support flight hours	48	591
Percent of total hours	25	75
Percent of cargo airlifted	48	52
M-60 tanks airlifted	19	-0-
M-48 tanks airlifted	10	-0-

Other cargos transported by the C-5 were two CH-53 helicopters, two A4E fuselages, 175mm cannons and 155mm howitzers. Fuel used by C-5s was 93 million pounds; fuel used by C-141's was 143 million pounds.

A high cargo-handling rate, termed "throughput" by the military, was necessary. Since ALCE had limited automated cargo handling equipment, the IDF provided muscular reservists and volunteer civil personnel, including teenagers, to move material. The results of these combined efforts completed unloading of aircraft within 60 to 100 minutes. This, coupled with the driving time to cargo depots, meant that from aircraft touchdown until depot delivery, the average time in Israel was three and one half-hours. By the fifth day of the airlift C-141 ground time was reduced to 55 minutes; for the C-5s the ground time was 100 minutes. During the last few days of the airlift, C-5s and C-141s were arriving in Israel with a frequency that made unloading times the critical element of the operation. This was emphasized by isolating a cargo B-747 for a period of days, until the airlift ended, prior to unloading the B-747. (Military crew notes.)

Of the 29 tanks, the C-5s delivered four to Israel before the cease-fire on 22 October 1973, the remainder being delivered within a week. Although the actual number is below 10% of battle losses totaling 432 tanks; the enormous psychological impact of main battle tanks motoring off a C-5A in pairs is better than lightning in a bottle. They proved the resolve of the United States to assist Israel in a very material manner. Moreover, this scene, featuring "tanks arriving by air" may have contributed to the armistice, sending as they did, the same signal to the Arab coalition. (21)

During the airlift, logistics reliability for the C-5 was 95.2 percent; the C-141 was 98 percent. A single C-5 was downloaded due to an abort. The timely delivery of these largest tanks was possible only with C-5s, an observation noted in several quarters. The General Accounting Office (GAO) summarized the capability in this manner: "The aerial delivery of combat tanks plus other outsize cargo by C-5s was an impressive use of airlift capability, and it is impossible to assess the full psychological impact of this demonstration"(22). The use of C-5s in this airlift, according to MAC's studies, "reversed the imbalance of military power created by the vast shipments of Russian war material to the Arab nations and led to a cease fire which in turn brought about a return to the earlier status quo. In short, airlift made possible the achievement of a national objective—peace in the Middle East" (23)

The depth of accomplishment of the American airlift was further accented by its difficulties. The lift was operated over narrow corridors on a 6,450-nautical-mile long route in international airspace. By comparison, the Soviet Union flew 935 missions to Egypt. Their distance was 1700 miles. This action delivered only 15,000 tons of material in 40 days. Thus the United States airlifted 25% more cargo, flew a route more than three times longer; and accomplished their objectives with 45% fewer of missions. (24) The C-5A proved its value during an unquestionably critical period of national interest.

The Miami Herald

C-5 Passed With Colors

Within nine hours after President Nixon decided on Oct. 13 to airlift military supplies to Israel the first U.S. Air Force cargo carrier arrived in Tel Aviv with a record 193,000 pounds of military supplies.

This was the beginning of a little noticed but truly historic operation dwarfing even the Berlin Airlift. "By Nov. 2," according to retired USAF Gen. Ira C. Eaker, "the United States had equalled the Soviet airlift to the Arabs, and by Nov. 22 it had doubled the Russian tonnage despite the fact that the U.S. supply line was 6,450 miles compared to 1,700 miles for the Soviets."

The airlift, which required only 566 missions to the Russians' 900 missions, doubtless saved an ally from extinction. "It restored Israel's weapons," Gen. Eaker goes on, averaging 1,000 tons per day for more than 30 days, "enabling her to regain the offensive, make decisive advances toward Damascus and Cairo, ultimately to achieve the ceasefire."

And how was this accomplished? Not only by superb pilots and crews but more to the point by a matchless airplane. We blush to report it was the much-maligned C-5A giant transport.

Along with many other newspapers, the Secretary of the Air Force itself, and countless congressmen, The Herald deplored the cost overruns and performance failures which appeared to make the C-5A the flying fraud of all time. In 1971 costs of the C-5A rose from $28.4 million per copy to $59 million, numerous wing and landing structure failures were reported, and the Air Force cut its order from 120 planes to 81.

But it worked. We wuz wrong.

Published December 1973.

AIRLAUNCH MINUTEMAN FROM C-5A, 8 OCTOBER 1974

The essential details of the flight testing required to accomplish the air launch of the Minuteman missile are discussed in an interview with Bill Harris, the Lockheed flight test engineer who wrote the test plan.

Q: Which C-5 flight test did you enjoy the most?

BILL: That is a tough question as there have been many interesting, exciting tests. But I would have to settle on the Air Mobile Feasibility Demonstration to air launch the Minuteman missile as the flight series that really stands out.

Q: The Minuteman missile is a heavy load to airdrop; what are the details of the tests?

BILL: I was in my third week of an assignment at Lockheed Missiles Division working on the Trident missile when the call came to join a newly formed team whose program was to air launch the Minuteman from the C-5. The tentative plan included receiving go-ahead by 14 August 1974, with completion of a successful missile drop by the end of October. The demonstration of air launch of the missile had to be complete prior to the Strategic Arms Limitation Treaty discussions being held near year-end of 1974. This was a monumental task to be completed in such a short time. It required the utmost in cooperation and teamwork of many organizations directed by Space and Missile Systems Office (SAMSO) of the Air Force Systems Command. Military agencies supporting this program included ASD, NPTR El Centro, SAMTEC, Vandenberg AFB, AFFTC Edwards AFB, and Hill AFB. Contractors included Boeing Aerospace Company, Space Vector Corporation and , of course, Lockheed Georgia.

Q: What are the principal elements in accomplishing this program goal?

BILL: Since the original C-5 airdrop program was completed at Fort Bragg, N.C. in 1971 engineering analysis was required to determine that the airplane was suitable for air-dropping a single package weighing in at 86,000 pounds. The airplane was originally designed for a maximum single package weight of 50,000 pounds and sequentially air dropping four 50,000-pound packages. However, during the course of the original test program at Ft. Bragg the Army indicated there was no requirement to airdrop any package weighing over 35,000 pounds. Since tests had already progressed to 42,000 pounds no further testing was accomplished. The maximum sequential airdrop in the original program was four packages with a total weight of 164,000 pounds.

For the current goal, it was necessary to establish test requirements and specify materials and equipment essential to accomplishing the task. Two instrumented C-5 airplanes would be required to meet the flight frequency and support other tests in the program. These ground and air tests were as follows: extraction parachute system development, airdrop weight build up tests, develop loading techniques for the special airdrop platform and missile cradle, ground EMI tests, vibration and dynamic response testing, and, ultimately, airdrop of inert and live ICBMs. Getting the ICBM off the airplane still required a method for separating the missile from the platform and cradle, and, finally, a parachute system that would stabilize the missile in the vertical attitude for proper ignition.

Q: What were the test conditions for these buildup flights?

BILL: All tests were performed at 20,000 feet, which in itself was a source of much of difficulty associated with the program, as far as the flight crew members were concerned. For example, most of the physical activity in the cargo compartment was done with the rear doors open; thus we were un-pressurized and sucking oxygen from standard oxygen walk-around bottles, normally good for 15 to 20 minutes. But the required physical activity depleted the bottle in 3 to 4 minutes. We switched to use high-pressure bottles on the later flights, thus a 20 pound package was added to the 40-pound parachute strapped to our back. In the C-5, the cargo compartment length dictated the use of a trailing inter-phone cord on this test cycle; the intercom plug is clipped to your flight jacket. I also used a movie camera during the tests.(We decided against coffee breaks until we got on the ground.) All parachute development testing and airdrops were done in a speed range of 160 to 170 knots.

Q: There must have been a couple of wrinkles in the sequence of testing, what were they?

BILL: The flights began with parachute tow tests to develop the extraction system. We began using a single 32-foot diameter parachute, but during the sixth dummy drop 86,000 pounds the extraction chute failed. This resulted in a decision to use a cluster of two 32-foot extraction chutes. Satisfactory development of the extraction system was followed by airdrops of increasing package weights from 45,000 pounds, in ten thousand-pound increments, to 85,000 pounds. These airdrops used a 46-foot platform, a special, minimum length, and designed to accommodate the 56-foot long Minuteman missile.

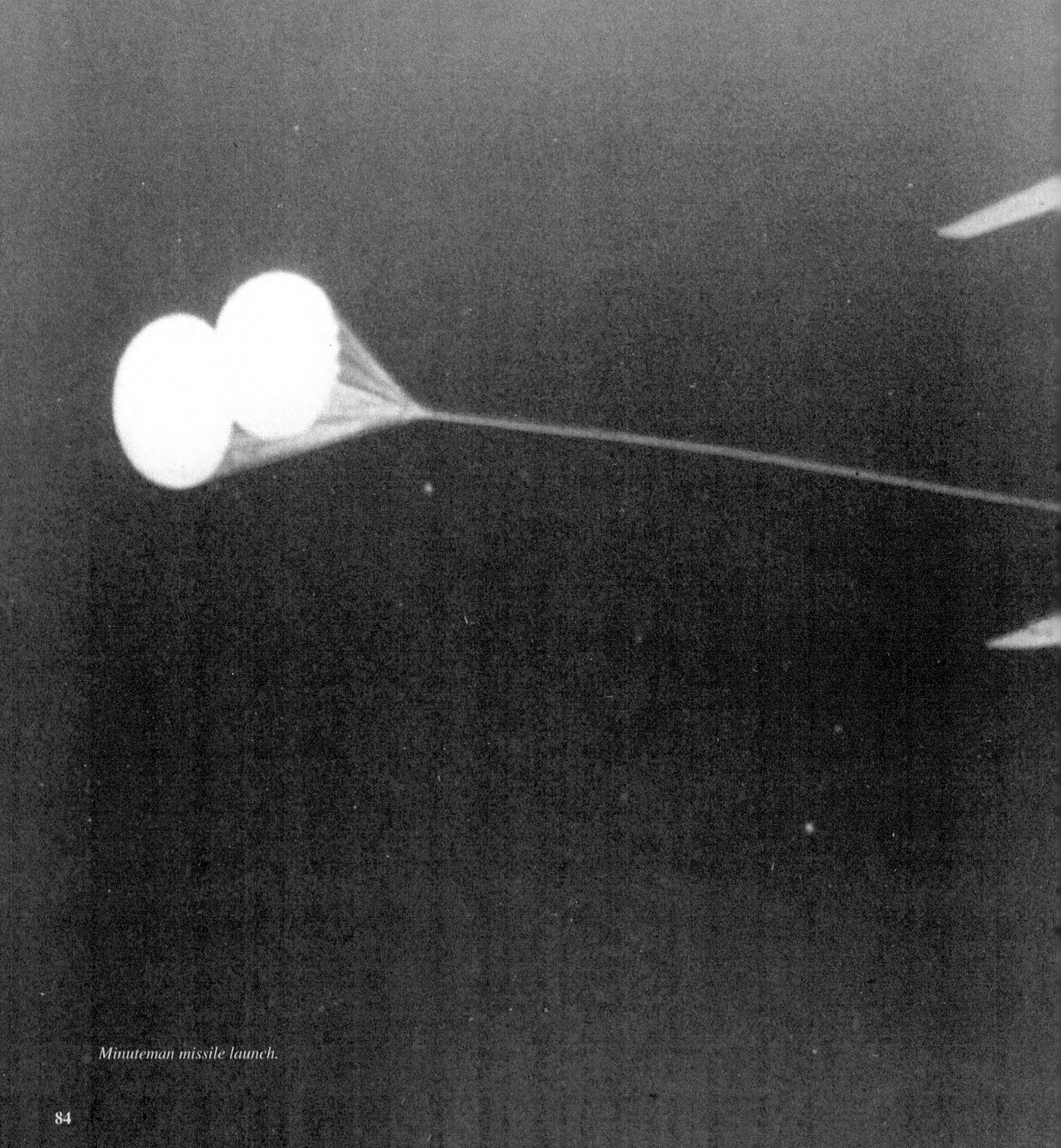

Minuteman missile launch.

The first of these two drop tests were similar to lighter weight packages; the exception was pronounced flexing of the platform where the test cradles butted on the platform. The third airdrop of 65,917 pounds was to be the last of the recovered loads; however, platform flexing was concurrent with improper recovery chute inflation. They streamed individually instead of simultaneously. The platform eventually failed and the test loads were lost. Since the recovery chutes were operating at or slightly beyond their design capacity the results were not a complete surprise. Because the weight of the remaining packages to be airdropped exceeded the capability of the K-loader, the Ballistic Missile Trailer, (BMT) was used for the remaining tests, which required development of special techniques to load the ICBMs. The height difference between the BMT trailer and the airplane cargo floor required the use of translating jacks to assure perfect alignment. A transit was used to provide the needed accuracy. Stabilization chutes were introduced for tests involving the dummy and live missile drops: another detail. After the fourth airdrop, the test tubs were welded together to minimize platform flexing. The 75, 000-pound airdrop was without the dreaded platform bending, but instrumentation failure required a repeat of this step in the progressive weight buildup. The next drop, a weight of 77,000 pounds, was successful.

The sixth airdrop, a package weighing 87,320 pounds, the heaviest single package ever airdropped, was the last test to use the single 32-foot extraction chute. Extraction chute failure on this event allowed the load to exit the airplane more slowly than normal, tip off the ramp, which in turn caused the stabilization parachutes to deploy too slowly. The test package tipping motion induced a violent oscillation, causing the load to disintegrate in mid-air. Despite the slow rate of deployment, the positive result of this event proved that the 87,000-pound package could be safely jettisoned from the C-5A. All future airdrops used a pair of 32-foot extraction chutes. The seventh airdrop, a dummy weight of 86,000 pounds, was successful in all stages, this cleared the way for the dummy missile airdrop test phase.

Q: Since there were problems during the parachute development and the dummy load airdrops, did this trend continue into the dummy missile and live missile airdrop phase?

BILL: As a matter of fact, there were no significant problems, we had three missile drops that were problem-free. While all the airdrop development tests were being completed, Boeing was at work designing and building a missile cradle compatible with the airdrop platform. This special support cradle had to fit the 42-foot pallet, provide support during the extraction event, and stabilize the missile through its transition to a vertical attitude for descent prior to ignition. The first inert package was airdropped at El Centro Test facility. It was virtually a perfect event. A very slight rotation of the missile was noted during descent. This test article was an "iron bird" with a guidance system active only to the level of providing system health and functional reference points during the descent to touchdown. The second inert missile, which was also loaded at Hill AFB, included inert first and third stages, with a live second stage minus the igniters. There were minor rigging changes to this missile. This airdrop occurred over the Vandenberg range off the California coast. Again the slight rotation of the missile was noted as the stabilization chutes deployed.

The third missile drop was the live article. This historic event took place on 24 October 1974, one week ahead of scheduled deadline. After takeoff from Hill AFB, Utah the C-5 flew a track to the Gaviota VorTac in Southern California, then into the Vandenberg drop zone. The airdrop was initiated at 20,000 feet at 160 knots EAS. The 85,324-pound missile extracted from the airplane normally, separated from the cradle and stabilized vertically in descent to approximately 8,000 feet where ignition occurred. The missile then flew to an altitude of about 22,000 feet during ten seconds of rocket propulsion. The inert missile then dropped into the ocean completing a very successful test and demonstration program

Q: Was consideration given to getting certification for an airdrop record?

BILL: Airdrop records are established for recovered packages only. We had planned to recover the 65,917-pound package, but the recovery chutes were operating at an overload condition and failed as they deployed, thereby preventing successful package recovery

Q: Does this program mark a career highlight for you?

BILL: There is a series of pluses experienced on a program like this, but the real salute goes to the exceptionally fine team of people who set new standards of performance daily to meet the objective of handing the "Russian Bear" a whole new set of problems.

Flight Crews Assigned to Air Mobile Feasibility Demonstration Program

	Primary Crew	Backup Crew
PILOT	*Jesse T. Allen	Carl A. Hughes
PILOT	Frank D. Hadden	Walt E. Hensleigh
PILOT	*Major Bruce Hinds USAF	Major Luck USAF
FE	*Jerry Edwards	Jack W. Parker
FE	Louie E. Lamb	Malcolm J. Davis
FTE	E.B. Brooks	Ernst Mittendorf
FTE	*Bill Harris	R.D. Edwards
TE	*H.J. Hunter	C.W. Marshall
LM-1	*E. Hardin MSGT USAF	J. Simms CMSGT USAF
LM-2	*T. Phillips, MSGT USAF	J.R. Beemer
LM-3	W. M. Parker	S Sgt. S. Storm, USAF

(*) These men were on all test and demonstration flights. B. Nagler and R. Scarboro, Boeing and Space Vector Company specialists, flew on initial tests.

From left: Hunter, Harris, Simms, Luck, Phillips, B. Edwards, Parker, J. Allen, J. Edwards, Hughes, Hardin. Elements of primary and backup flight crews for C-5A Minuteman Missle Air Launch.

Loading dock for a C-5.

C-5 refueling from a KC-135 tanker.

CHAPTER 8

RECORDS WERE MADE; THEY WILL BE BROKEN.

C-5 Galaxy World and National Records

World and National Records Certified by the Federation Aeronautique International (FAI) and the National Aeronautical Association (NAA)

On 17 December 1984, at Dobbins AFB, Marietta, Ga. one world aviation record and one national aviation record was set by the C-5 Galaxy.

*World record: "greatest payload lifted to 2,000 meters, a total of 245,731 pounds.

*National record: "the greatest recorded weight any aircraft has ever flown", a gross weight of 920,836 pounds.

On 16 January 1985, at Edwards AFB, the C-5 set a national record for the heaviest aircraft ever landed. The landing weight was 876,762 pounds.

The above records were set during testing of the C-5, which set the following additional Company records.

*MAXIMUM TAKEOFF WEIGHT	841,571 POUNDS
*MAXIMUM INFLIGHT WEIGHT	922,200 POUNDS
*MAXIMUM LANDING WEIGHT	876,762 POUNDS
*MAXIMUM BACKING WEIGHT	770,700 POUNDS

On 12 July 1989 at Pope AFB N.C. a USAF aircrew from Dover AFB air-dropped four 42,000 pound Sheridan tanks and 73 paratroopers. The total weight is 190,396 pounds. This event was part of the AIRLIFT RODEO 89. Representatives of the National Aeronautical Association were on hand to authenticate the equipment weights and validate the airdrop procedures. The airdrop was completed in two passes; the four tanks were sequentially airdropped on the first pass, six minutes later the 73 paratroops jumped.

The NAA has petitioned the FAI to create a new category and certify the airdrop as a world record.

The C-5A airplane has one unofficial record in the annals of airplane procurement. It posted this record in December 1969 with the delivery of the first C-5A to Altus AFB. In the time interval, 1945 to 1969, it is the only aircraft purchased by the military that has met or bettered every performance guarantee in its contract. The data is presented tabulated on the following page.

C-5A Contractual Guarantees and Performance Results

These program achievements were demonstrated by Category I and Category II Engineering flight, flown by both contractor and military test pilots.

	Cargo weight	Guarantee	Actual	% of Guarantee
PEACETIME MISSIONS	112,000 LBS	5500 NM	5583 NM	102%
	100,000 LBS	5800 NM	5906 NM	102%
	200,000 LBS	3450 NM	3519 NM	102%
	220,000 LBS	3050 NM	3086 NM	101%
WARTIME MISSIONS	265,000 LBS	2700 NM	2764 NM	102%
	265,000 LBS	2500 NM	2666 NM	107%
	200,000 LBS	4,000 NM	4123 NM	103%
ADDED MISSIONS	100,000 LBS	5500 NM	5667 NM	103%
	200,000 LBS	3100 NM	3278 NM	106%
	200,000 LBS	2700 NM	3096 NM	115%
	220,000 LBS	2700 NM	2844 NM	105%

	Distance	Speed	Cargo Weights	Actual Weights	Guarantee
RESUPPLY MISSIONS (in/out)	2500/2500 NM	@440kts	100,000 LBS in/ 0 out	114,790 LBS in/ 0 out	115%
	2500/2500 NM	@460kts	100,000 LBS in/ 0 out	121,080 LBS in/ 0 out	121%
	4000/1000 NM	@440kts	117,800 LBS in/ 0 out	134/450 LBS in/ 0 out	114%

The C-5 bettered by margins of 1% to 21% the fourteen contractual mission guarantees.

Takeoff and Landing Guarantees vs. Demonstrated Results

	Guarantee	Actual	% of Guarantee
728,000lbs T.O.W.	8,000 ft	7,620 ft	105%
769,000lbs T.O.W.	9,100 ft	8,570 ft	106%
OTHER MISSIONS	7,500 ft	7,500 ft	100%
RESUPPLY			
A-LANDING	4,000 ft	3,300 ft	117 %
B-LANDING	4,000 ft	3,360 ft	116 %
C-LANDING	3,900 ft	3,225 ft	117%

These take-off and landing performance results meet or exceed all guarantees.

C-5B cockpit

CHAPTER 9

"THE C-5B IS A MUCH BETTER AIRPLANE, MORE BEEF, FEWER MOVING PARTS."

C-5B, An Internal Maturity for the Galaxy

H. Bard Allison

Executive Director, C-5/C-141 Programs 1983 to 1987

Q: Can you outline a few of the more obvious differences in the C-5A and C-5B business environments?

BARD: The C-5B program had a series of advantages. One that was a very big factor, yet overlooked because it was not pointedly obvious, was the state of the C-5A spares inventory. When the "A" production shutdown came in the early seventies, the spares pipeline and inventory declined rapidly. When President Reagan took office in 1982, his "kitchen cabinet" made him aware that the military was fast depleting its war emergency spares. Thus, a program to purchase spares was put in place. Major items like engine turbines, actuators, wing leading edge slats and the like were procured. In addition, AVCO in Nashville, Tennessee was producing new wings for the C-5A fleet retrofit. This supplier activity actually put new life in the vendor capability. As a result the parts procurement risk was minimumized when the C-5B contract was in work later on. Another big plus was the manufacturing manpower technical skill level that Lockheed Georgia had taken steps to develop. At a location on Franklin Road in Marietta, Lockheed ran a training school with workshop where potential new hires could learn, develop, and demonstrate sheet metal skills. This schooling was on an individual's own time, a non-pay workshop where quality techniques were acquired and demonstrated prior to employment

Q: Those are very big factors; how did others come into play?

BARD: The list is a bit long; we had on board an engineering and manufacturing team that had been through the tough days of the 60s when the "A" was developed and built; plus they had just recently completed the stretch work on the C-141B, a remarkably sound program performance in every category. Also, the company was in the middle of the C-5A wing mod program which was showing excellent results. These facts, and the availability of over 90% of the basic manufacturing tooling, made the company position very strong. We proved to the customer that the risk factors on the "B" program were minimal.

Q: The C-5A always had a reputation for being very difficult to maintain; what is new in the "B" to eliminate some of the problems?

BARD: A substantial portion of that "A" poor maintenance reputation is self induced, if one recalls those years where spares were not being purchased. What happens if spares are not available is that one airplane becomes a parts bin, and operative equipment is "cannibalized" to keep other airplanes operational. Now there is an operational necessity for this procedure, but, the distortion of facts on maintenance man-hours (MMH) comes about where the parts removal and reinstallation hours are charged as maintenace hours, distorting the whole picture. In addition, the aircraft utilization was down to 1.5 hours a day so the tow and wash-rack time also distorted the numbers as a result of low airplane flight time. Back to your question: several changes in the "B" make the airplane much more maintenance friendly. Removing the Crosswind Gear function from the airplane let engineering simplify the mechanical portions of the main-landing-gear-door system and this type ofsimplification always reduces the man hour expenditure. We replaced the beryllium brakes discs with carbon discs, a very important change that removes a special-handling material from the maintenance inventory, a positive in two sectors. Basic material for structure, such as wing panels, mainframe forgings, frame straps, fuselage under-floor end fittings, wing/pylon attach fittings and the aft ramp lock hooks were changed to materials that had much better corrosion resistance characteristics than the materials used in the "A". Each of these changes heavily reduces the major maintenance requirements. All changes were subject to a thorough Air Force and Lockheed review team examination which took special note of the operational and maintenance history of the C-5A.

Q: Sounds like the customer was going to be very happy to get the new Galaxies.

BARD: The politics of aircraft buying creates some strange circumstances. There was a significant group of Air Force and procurement people that insisted the "B" was a high-risk investment for the government. USAF people who were in the mood for a new cockpit, somewhat akin to the new car every five years theory, showed little concern for factual information. The C-17 was in development, and we had to contend with the unbridled enthusiasm of several customer groups for a machine more than twice the cost of the "B" that in critical conditions airlifts only half the payload. Happy customer? Not exactly. Yet the Air Force personnel at the plant in Marietta worked hard to ensure a quality product to the user. Only Desert Storm put the true picture in focus. The "B" really performed very well during Desert Shield and Desert Storm

Q: There were numerous changes to the C-5B internally, but the one that gets a lot of discussion is the deletion of the crosswind landing gear; why did that get taken out?

BARD: The maintenance organizations of the MAC community made an irresistable case for removal. Discussions went on with the ops guys, naturally, but the airplane history could not show that CWS was being used very much, if at all. Pilots like their toys, but with the way the airplane is presently operating, CWS is not necessary. The changes in a large number of the avionics, such as better radios, interphone, MADAR, color radar, inertial navigation systems and the Fuel Savings Advisory System (FSAS), are primarily due to the avionics industry advances in their equipment.

Q: This program was a fixed price contract with the government, but some accounts indicate that the company was squeezed at the end of the program. What happened?

BARD: The fact of the matter is very straight forward. The contract contained a fixed price for fifty airplanes, broken down into fiscal year buys. The first buy was only for five airplanes. We produced, the customer liked what they saw, and so continued buying through the fiftieth machine. The government always has the leverage to renegotiate a fiscal year buy which they did on the later procurements. Details of the yearly contracts changed as the experience ledger was reviewed, and both the Air Force and company finished a successful program with high marks for quality of the C-5B.

Now with that positive accomplishment in the record, do not think that the program was without a few hair-pulling sessions. The manufacturing plan called for increasing production from one to two C-5s a month. This caused a major ripple in getting factory functions working smoothly. Those fiscal year contract renewals were not without hassels. The final one, for the last 21- airplane buy, required Lockheed's corporate president to settle pricing with the Air Force general in charge of procurement.

Q: What was your experience on the "A" program?

BARD: During that program I directed the engine integration into the airplane. Developing both the engine and the airplane simultaneously was a real challenge. Our engine supplier, General Electric, was in an interesting position. They had their own separate contract with the Air Force, but Lockheed had "Total System Responsibility". This caused less than smooth start up negotiations early in the program between GE and Lockheed. The administrative routine called for monthly meetings between the two of us, bringing your problems to the table and hammering them out. In one of these face-offs with at least ten experts on either team, the conversation became very heated. Tom May, our executive vice president and program manager was chairing the Lockheed group, along with Gerhard Newman from GE. Tom chewed about 30 millimeters out of his cigar while taking in the verbal sparring, turned to Gerhard and said, "Let's go to your office." When they returned, Tom sat for about ten seconds, eyeballed everybody, then stated. "Gerhard and I think this contract to build the C-5 with GE's turbo-fan power is mutually beneficial to both companies. We plan to work in that direction. We have decided that the best engineering solutions will be the least cost to both parties." We made real progress from that moment on since it let the engineers get on with the job at hand.

Q: The TF-39 is a huge departure from the turbo jets of that time, who were the in house engineers who kept a good grip on the engine development and installation progress?

BARD: It was a team effort with the project design team at Rohr Corporation in Chula Vista, California and the aero-thermo and installation team in Marietta working very closely with the General Electric group

Q: Did you get much of an opportunity to fly on the airplane?

BARD: Yes, on both "A" and "B". I was on the delivery flight for the first "B" to Altus AFB. It's always fun to get an opportunity to visit the customer where the rubber hits the runway.

C-5 Configuration Changes From A To B

The C-5A and C-5B have the same external dimensions, use the same engines, fly the same takeoff, cruise, and landing speeds, but there are substantial internal system differences between the two airplane configurations. The "B" configuration of the GALAXY benefited from many major and minor improvements in system designs, manufacturing techniques, materials processing advances that had evolved since the original C-5A production. Also, the education that accrued during eleven years of actual operational experience was utilized to influence the C-5B configuration design. The C-5A production line ended with the delivery of the last "A" airplane in May 1973. The "B" production award was made in December of 1982, almost ten years later. The first C-5B flew in September of 1985. In addition, and most importantly, continuous use of the C-5A over this time interval certified a number of important airplane characteristics which could only be identified in actual flight operations. Heading this list of positives were the following:

1) The superior flying qualities of the aircraft in the takeoff and landing configuration, especially under adverse weather conditions.
2) The engine reliability experienced under a variety of operational conditions.
3) The engine tolerance to uneducated or unintentional abuse.

Thus, a large number of system improvements were mandated in the production of the "B" airplane. Using C-5A, ship 0081 (the last production C-5A), as a basis, the following is the list of production changes in the "B" airplane

New Wing Design

The most significant difference between the "A and B" was the wing structure. The C-5B wing structure is identical to the modified wing that was installed in all 77 remaining C-5A airplanes. Details of the reasons for wing redesign are found in Chapter 4. After extensive study by Lockheed, USAF, and independent blue ribbon agencies, a redesign of the wing structure was accomplished. Comprehensive testing followed, putting the test specimens through 105,000 cyclic test hours, plus prototype flight tests. A wing modification contract was signed between USAF and Lockheed Georgia, which dictated replacement of the wings of all C- 5A airplanes. The "A" model airplanes were cycled through the Marietta facility for new wing installation. This comprehensive modification program began with the approval of the new wing concept by the USAF Air Council on 10 August 1977. The design was completed June, 1978; prototype testing began in August of 1979. The initial C-5A was modified beginning in January 1982; the final C-5A airplane new wing structure installation completed in May 1987.

The new stronger wing reveals itself in subtle ways. Flight performance does not change, and, although pilots report minor detectable differences in maneuvering, or dynamic, characteristics due to the stiffness change, it is in the steady state or static situation where the difference is pronounced. Wing deflection on the ground with a full fuel load is substantial in the unmodified C-5A, with the 332,500 pounds of fuel load defection reduced to 1.5 feet, which increases wing tip ground clearance to 14 feet, 8 inches.

The modified wing installation process itself was unique in a number of manufacturing techniques utilized. The engines, engine pylons, flight controls, flaps, slats, spoilers, hydraulic actuators for each of the aerodynamic surfaces, the fuel tank pumps, and plumbing components of the hydraulic oil coolers, were all removed and stored for replacement on the new wing. Once removals were complete, the old wing panels were literally sawed off—using a chainsaw.

Cross-Wind Landing Gear

The second most notable difference for the C-5B is the deletion of the cross wind landing gear system. In the early 1960's when the Air force planners conceived the C-5 airplane, flight control of large airplanes in average crosswind conditions was especially difficult. The addition of crosswind gear capability was included in the original specification for the C-5. The utility of a cross-wind landing gear allows the fuselage/wing to face the prevailing wind while the wheels (landing gear) point down the runway; thus full aerodynamic flight control remains available to complete the take-off or landing approach. In the 1960s, this requirement made sense for a military transport whose cargo must arrive on time in combat scenarios. Operational experience with the C-5 in the 1970s, which included the Vietnam conflict, led the Air force and Lockheed to the conclusion that flight control of the airplane was quite sufficient, even in substantial crosswinds. Thus, the cross-wind gear has been eliminated from the "B", and will be eliminated from certain "A" airplanes in the future. Flight operations into and out of fields with 90 degree crosswinds up to 28 knots can be accomplished safely and routinely. This is not to suggest that takeoffs or landings in high crosswinds are simple, or do not require pilot focus and skill. The point to be emphasized here is, more than adequate control capability exists in the airplane.

Increased Engine Thrust

The third major change for the "B" model is the increase in engine thrust. The General Electric engine, designated the TF39-GE-1C, produces 5.2 percent more thrust than the -1A version, the original engine powering the C-5A. One additional consideration of engine matters: had the C5-B never been produced, the improved -1C version of the engine would still have been developed, produced and utilized by the existing C-5A fleet.

Other detailed improvements which were incorporated in the C-5B are shown below.

Improvements to the C-5 Airplane in 1985

Wing-Empennage-Pylon

a) Zero fuel weight increased 20,000 pounds.
b) Wing/pylon attach fittings replaced by more corrosion resistant steel (PH13-8Mo).
c) Access to nitrogen dewars improved.
d) Aft engine mount material in pylon changed from H-11 to Inconel 718 grade.
e) Near net forgings used in lieu of machined fittings for flap carriage rib fittings.
f) Hi-Lock fasteners replaced with Hi-torque titanium in the empennage.
g) Seven additional manufacturing changes.

General Structural

a) Cargo winch changed to hydraulic drive from electric drive.
b) Four maintainability requirements addressed.

Fuselage

a) Nose landing gear chem milled skins replaced with standard production skins.
b) Aft upper lobe skins, FS 1523 to FS 1964, increased in thickness.
c) Aft pressure door backup fittings redesigned per ECP 6685.
d) Flight station and courier compartment floors redesigned.
e) New corrosion protection; six production changes.
f) Thirty-two added manufacturing and design improvements.

Propulsion-Nacelle-Fuel System-Fire Protection

a) Complete liquid-nitrogen fed portions of the fire suppression system installed. (This system was added to the C-5A in 1975.)
b) The optical fire detection functions of the FE-1301 system were installed, added to C-5A in 1975.
c) Modified fuel vent system installed.
d) Engine nacelle blow out doors modified with hinges.
e) Non-clogging hydraulic pump case drain filters added.
f) Air turbine motor (ATM) changed, improved reliability.
g) Eleven GE-TF-39 engine component improvements.
h) Sixteen additional system changes.

Flight Controls-Mechanical Systems

a) Ten component improvements.

Avionics

a) Interphone system replaced by AN/AIC-18 system.

b) A Fuel Savings Advisory System(FSAS), a computerized flight planning and tracking tool, installed in the center instrument console.
c) Latest design of short wave radio communication, HF/SSB, and the improved IFF gear (APX-100), installed.
d) Thirteen added communication/navigation wiring changes.

Instrumentation

a) A redesigned Malfunction Detection Analysis and Recording subsystem(MADAR). This device allows the flight engineer to monitor principal airplane subsystems, including the engine. MADAR includes a capability to dynamically balance the TF-39 front fan, eliminating the necessity for engine removal for an out-of-balance repair.
b) Delete the integral weight and balance system (IBWS).
c) Replace the ejectable Crash Data Position Indicator (CDPIR) with a crash survivable unit.
d) Install loads environment spectrum survey device (L/ESS). Ten percent of the C-5B airplanes equipped with this instrumentation.
e) Four additional instrumentation requirements.

Electrical System

a) Electrical power system controllers replaced with current level technology.
b) Windshield heat transformers relocated.
c) Nine additional electrical system changes.

Hydraulic System

a) Replace Mil-H-5606 fluid with higher temperature Mil-H-83282 fluid.
b) Use 21-6-9 steel lines instead of AM-350 steel.
c) Engine driven hydraulic pumps were changed to the redesigned Vickers or Abex pumps.
d) Strength of hydraulic manifolds were increased, interchangability with C5-A systems was retained.
e) Seventeen additional hydraulic system changes.

Main Landing Gear

a) Carbon disc brakes installed in lieu of beryllium brakes. Both the "B" brakes and the "A" brakes are inter-changeable.
b) Simplified main landing gear(MLG) door actuation system removing redundant linkage and actuators.
c) Deletion of the cross wind landing gear function allowed fourteen addition redundancies.

Enviromental System

a) Five system changes.

Automatic Flight Controls

a) Redesigned the Go-Around Computer, the stallimiter computer, and the control wheel hub force sensor. Retained interchangability between airplane configurations
b) The yaw/lat computer, pitch augmentation computer, the auto throttle computer, and the automatic flight control panel were upgraded.
c) Four additional changes.

Flight Station Instruments

a) The crosswind gear controller and indicator were deleted.
b) New engine instrument reflecting the use of fan RPM as primary power indicator. The engine pressure ratio (EPR) gage was eliminated from the panel.
c) Eighteen controller/instrument revisions in the cockpit.

In 2001, DOD has contracted with Lockheed, Honeywell and General Electric to provide performance improvements for the C-5 airplanes. Honeywell will supply integrated avionics systems which combines avionics and navigation equipment with liquid crystal cockpit displays for all engine and com/nav data. GE will provide the CF6-8C2 engines. This engine will provide 22% increase in thrust, improving initial cruise altitude from 24,000 to 33,000 feet while reducing takeoff roll and time to climb. Lockheed Martin will be the program manager for the modifications.

C-5B first liftoff, 10 September, 1985.
(3) Pilots: Dvorscak, Narleski, Hadden
(3) FEs: Edwards, Garger, Scott
(3) FTEs: Young, Jennings, Sutherland

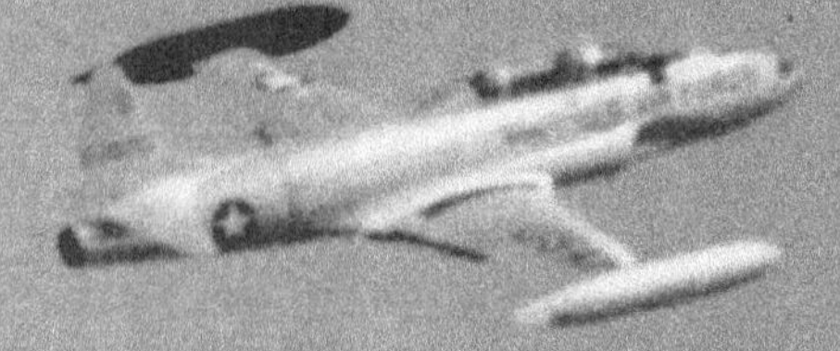

C-5A with T-33 chase.
Vern Peterson chase pilot

CHAPTER 10

VIVID MEMORIES FROM LOCKHEED EXECUTIVES
REMARKABLE FLIGHTS OF SKILLED PILOTS

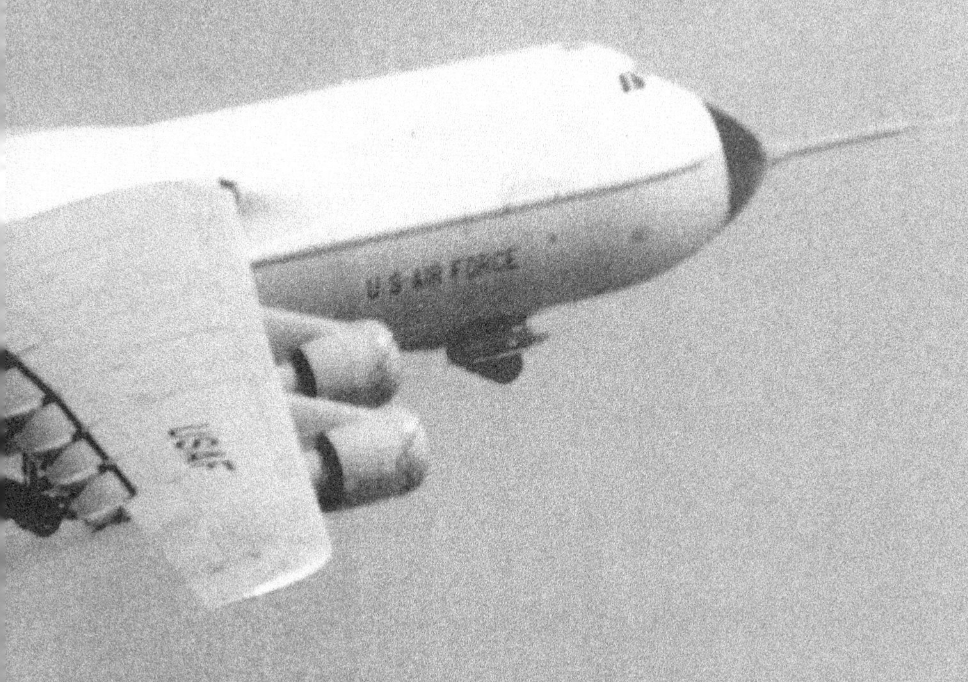

INTERVIEWS

Details of the development of the C-5A airplane, its follow-on production and final release for operational use by the USAF can be told by a recitation of the numbers that are integral to every phase of this contract. Much of that type of history has been written in the earlier chapters.

This following short series of interviews will detail special personal experiences recording individual and group actions, which in the immediate locus of work, presented difficulties requiring "inventive persistence" to overcome. The rapid resolution of these problems, often parceled out as challenges, was absolutely vital to the completion of the very ambitious program defined as the C-5 project.

Biographical data on each person interviewed is listed in the appendix.

E.B. GIBSON

C-5A ENGINEERING PROGRAM MANAGER

BACKGROUND

Lockheed operated the AF Plant #6 at Dobbins AFB in Marietta, Ga. for 14 years prior to the initiation of the C-5 production and development program. In that time period a B-29 and B-47 modification facility was progressively developed into a full-scale aircraft engineering development and manufacturing facility. When the YC-130 program was transferred from Lockheed Burbank to the Marietta facility in 1954 it set the stage for developing an outstanding engineering staff, as well as substantive upgrading of manufacturing facilities.

Following on the success of the C-130 Hercules program was the equally successful C-141 Starlifter jet transport which aided in further development of the engineering talent base of the Georgia facility.

THE COMPETITION FOR THE C-5A CONTRACT

The technical requirements presented by the USAF team were presented in an outstanding manner. Although the requirements were very challenging, the documentation presented a very balanced statement of facts. Lockheed's competitive submittal met or bettered every requirement utilizing the then advanced, but realistic, state of the art. The design margins were adequate, from 2 to 5%, in all technical areas.

The design "measure of merit" was minimum total system lifetime costs once all formal system requirements are satisfied. Both Lockheed and Douglas religiously exercised this absolutely correct design approach. Boeing's approach was very optimistic in their weight estimate, intentionally or unwittingly. They, among other details, decided to offer a higher than required, greater than optimum cruise speed in their proposal. This competitive ploy strongly influenced the final product specification once the contract was awarded.

C-5A PROGRAM REQUIREMENTS

The formal C-5A program competition was conducted in late 1963 and 1964. A wartime scenario involving a major conflict with the Soviets and parallel minor requirements elsewhere were used to establish and size the outsize airlifters.

Chapters 1 and 2 detail the payloads, performance and other systems requirements that were prime to meeting the specifications.

Total Package Procurement Contract (TPPC) elements had overriding impacts. DOD made ambitious efforts to capture a price regardless of the industrial market conditions prevailing over the life span of development and production of the airplane. The top management of all three competitors, Lockheed, Douglas and Boeing, knew the lack of realism in the TPPC terms, but each, in their own manner, planned to renegotiate as the specifics of marketplace pricing became crystal clear once production was ongoing.

The empty weight of the airplane was the one number in the contract "set in concrete." As final design proved, the need to increase the empty weight could be managed without performance penalty. General Electric had provided for a thrust increase in the engine, for a price. But the empty weight limit was not changed. Thus a costly weight reduction effort was the only way to prevent voracious penalty clauses from taking effect. The punishing effect of the TPPC contract was avoided by Lockheed's suppliers by the simple ruse of arbitrarily rescheduling deliveries, another cost impact that Lockheed had to fund.

One particularly unbalanced contract provision stated should Lockheed fail to meet any contract provision, USAF could force Lockheed to turn the contract over to another contractor of the Air Force's choice, with Lockheed being liable for any costs incurred above the original price.

Technical Level of Effort

Lockheed's study effort for the heavy lift system (HLS), which preceded the C-5A competition, involved several hundred personnel. In the five months prior to contract award the competition staff peaked at 800 people working seven days a week. This was, to this time, Lockheed's largest, most intensive, competitive effort, and a successful one.

The C-5A engineering design team's design, development and test program employed over 6,000 technical personnel. More than half of the team was non-Lockheed personnel. Two thousand, plus, were located off base in England, Canada, and the US. All supervisory personnel were Lockheed employees, many of whom had worked together on the C-130 and C-141 programs.

Engineering Budget Estimates

The Lockheed engineering organization submitted a budget that was developed using recent experience from the C-141 program, and factoring into the budget, allowances for the much greater level of effort for design challenges imposed by the complex systems which made this airplane, the world's largest military cargo carrier.

The finance branch completed a parallel study using some previous C-141 data plus statistical "ARCO" factors, which resulted in a less than reasonable comparison of "apples and oranges". The man-hours budget provided to engineering after contract award was less than 50% of the figure submitted to top management. This pressure point was severely increased as the magnitude of the weight control measures was brought to bear. The actual man-hours expended on the total program numerically validated the original engineering man-hour estimate.

Technical Submittal

A single copy of the formal technical submittal of 8"x11" bound documents covered a 40 foot shelf. It did, however, indicate Lockheed's complete awareness of the airplane specific requirements and WADC's technical material and process requirements. The aerodynamic configuration was defined in very great detail. Every functional system and operational component was documented. Both estimated and guaranteed performance was presented, indicating how the C-5A met or exceeded the requirement.

In this most important area Lockheed's advanced design organization had the established practice of first estimating achievable performance, back off from this point a small percentage to provide the guaranteed performance a positive margin. The percentages varied from 2% to 5% depending on the precision with which the calculations could be performed.

In final discussions between USAF and Lockheed top management, without the benefit of engineering expertise from either side, these internal company design margins became the performance guarantees prior to contract award. Thus another pressure point was enabled in the devilish TPPC contract.

Last Minute Negotiations Produced Costly Changes

In the final negotiations USAF has the advantage of culling all three proposals for "good ideas" and inserting them as a reconfigured format into the last minute discussions.

This happened "in spades."

First, recall that Boeing's weight estimates were overly optimistic. As evidence of Boeing's poor capability in the 1960's in this technical area, Boeing's first 747 exceeded its originally estimated empty weight by over forty thousand (40,000) pounds.

Secondly, Boeing's proposal offered a cruise speed higher than required, i.e. M=0.82. All earlier studies of the HLS systems, plus the independent Rand Corporation studies, concluded that the optimum speed for long range cruise is M=0.767.

Thirdly, the Total Package Procurement Contract required pricing on all aspects of the design, development, and spares for the life of the aircraft. This type of forecast on prices required discarding common sense, replacing it with guesswork. This gross demand became increasingly magnified as the contract aged.

These conditions led the Air Force, which preferred the higher speed Boeing design, but were faced with a Lockheed design which equaled or bettered every requirement at a total program price substantially lower than Boeing's, to demand that Lockheed meet Boeing's weight estimates. This demand imposed a 3.8% decrease in Lockheed's empty weight guarantee, which was already aggressive. A reduced empty weight was possible only by improving the weight state of the art in every design area, achievable only at great cost in redesign effort and man-hours.

Following the technical presentations and discussions, during which complete communication existed between Lockheed and WADC (Air Force) technical personnel, the USAF director of WADC, General Terhune, asked Dan Haughton, Lockheed's Chairman of the Board and a limited number of WADC top level non-technical persons to a private meeting. In this meeting the flight performance numbers, which exceeded guarantees and were in fact design margins, were innocently assumed to be the guarantee numbers by both parties. The resulting Terhune and Haughton agreement set in concrete another technical challenge beyond the original specification. While WADC technically astute personnel later admitted this erroneous conclusion on the part of top level management, DOD's Under-Secretary for Defense, Harold Brown, maintained a "hold their feet to the fire" policy. This program error added millions to the final bill.

Perhaps the most important lesson learned from the C-5 program is that top leaders making decisions on strictly technical issues need to have the qualified background or include qualified personnel in final decision issues. This trend is ongoing. More and more Chief Operating Officers are experienced engineers.

The actions noted here cost both Lockheed and the US government many millions of dollars since enormous added effort to achieve substantial advances in the state of the art were necessary to achieve the higher technical goals demanded.

During extensive joint USAF-Lockheed test programs the C-5A met or bettered every technical requirement and performance guarantee guarantees based on earlier estimated performance. When accepted by the USAF the C-5A was the largest and most complex aerospace vehicle ever procured for the military. It was the only new USAF aircraft ever to demonstrate such complete compliance during its acceptance-testing program.

It is appropriate to mention the issue of the C-5A wing. The test article in a fatigue test environment, "apparently failed" prior to meeting its operational fatigue lifetime goal. This "goal" was not contractually a guarantee.

There was a later decision to rework all C-5A wings as a result of the "possible" fatigue life problems. This concern was based on the original operational set of scenarios which included a terrain following mission, which was soon abandoned, however it was part of the fatigue article test series. Lack of adequate quality control on early wing structures including, in particular, the fatigue article itself was an additional factor. This area of speculation is further discussed under "lessons learned."

The Engineering Program

Organization

The organizational structure most adaptable to this program is a bilateral organization. In this arrangement each engineering program manager has his own functional engineering group staffed by specialists assigned from their parent organization.

The entire engineering organization, both the independent technical groups and the program groups reported to a single director of engineering.

The director of engineering's responsibility is to build and maintain an adequate engineering technical organization from which to provide each of his program managers the technical staff and support necessary to his program.

Individual program managers direct functions, and control budgets that relate to their program. Managers of the technical organizations are responsible for personnel performance and for quality of the technical data developed in support of each program.

Team Buildup

Prior to the formal C-5A competitive effort, Lockheed's level of effort involved about 200 to 230 engineers. The invitation to compete for the C-5A design, development, test, production and support program quickly increased this group to over 800 engineers.

During this effort USAF required the application of the 375-5 program management directives, which required Lockheed to employ a large group of systems engineers. Use of the 375-5 directive series had a number of positive benefits, but its use also created many issues which were a net negative in an environment of daily schedule pressure.

At its peak, Lockheed's technical team reached a level of 6000 engineers working an average of 48 hours a week, an equivalent of 7,200 engineers – by far the largest aircraft technical team ever assembled to date This rapid and successful buildup of a superb technical team surpassed an earlier industry effort and even Lockheed's initial expectations. Credit is due to the fine cooperation of the aircraft industry, including both of Lockheed's C-5A competitors. Of particular value was Douglas's cooperation. That corporation gave Lockheed a copy of their complete C-5A competition technical submittal.

Lockheed hired 1500 engineers as permanent employees, and a number of subcontractors provided technical personnel on loan. Several British firms provided engineering talent for the 1,000-man effort that accomplished the London-based wing design support under Lockheed Georgia guidance.

Policies – Practices

To assure up-to-date recognition of progress vs. requirements or guarantees, Lockheed employed the "Pert-Tech" system. This method, using computer programs, related the current estimates of the variables, weight, lift, drag, installed engine thrust, and other factors which would impact the performance guarantees. Monthly, the best current estimates of the state of progress vs. every critical performance-related technical factor was run through the computer program to define progress toward meeting requirements.

The output of these programs produced positive or negative margins versus the performance guarantees, including the impact variations in the value of factor inputs would have on each guarantee. These computer data allowed technical emphasis where maximum benefit could be derived.

Weight and drag control, particularly in second segment takeoff climb, became the greatest technical challenge. The second segment climb drag, combined with the higher than anticipated thrust lapse rate of the high-by- pass ratio turbofan engines, resulted in major redesign of the high lift system. This redesign was accomplished, with the outstanding work of the London based engineering staff, very successfully. Thus the final ratio of drag vs. lift in the second segment climb configuration was by far the best ever achieved for a modern jet transport airplane.

This major design control methodology was pivotal in allowing Lockheed to meet each performance guarantee. In addition it also helped create trust between Lockheed's technical management team and the C-5A WADC SPO. Internal and vital Lockheed "(Pert-Tech)" reports were passed on, without editing, to the C-5A SPO technical director, Bernie Lowery, and his boss, the SPO director, General Guy Townsend.

Schedule Management

Contract management required PERT-TIME format to control the schedule. However, the company employed a much more detailed, disciplined, yet flexible schedule negotiation methodology to manage the thousands of related details.

This management aid:

A. Identified intra and inter organizational schedule interfaces in the computer program, initially over 200,000 events in all.
B. The individual discipline was allowed freedom of rescheduling; inter discipline rescheduling was tightly controlled.
C. The program produced daily schedule performance and highlighted summary reports for every level of management, from the lowest tier to the top.
D. For schedule and reschedule negotiations to be officially accepted, the proposed change in schedule was entered into the computer program by schedule monitors assigned to assume all affected managers participated.
E. Schedule impasses, when occurring, were thus quickly brought to the appropriate senior manager, including the C-5A engineering program manager.
F. Seldom did an impasse require more than one level up in management to resolve.

USAF Relationships

Lockheed's relationship with both the WADC organization at Dayton and DOD in the Pentagon had an established history originating in the C-130 and C –141 programs. These business encounters were, in most cases, mutually trusting and cordial. However, a degree of skepticism was introduced by Colonel Townsend, later head of the C-5A SPO, whose experience at Edwards AFB with the C-141 test program resulted in significant lack of optimism for the Lockheed-run C-5A program.

At the outset of the contract the engineering policy was established to be always open, candid , and most importantly, honest with USAF counterparts at every level. This policy, including the use of the USAF 375-5 program management directives paid real dividends. The unedited PERT tracking data passed from my office intact to the SPO counterpart, Mr. Bernie Lowery.

SPO conservatism did override our earnest projections, resulting in a "cure" notice. This official report of serious concern with the program progress resulted in a joint USAF, industry, and academic review team checking all aspects of the C-5A engineering program. After a several week study the program made no changes at all. Not a single rivet, bolt or nut was changed. Undoubtedly, the reviewing team learned a great deal about the technical advancements in state of the art for large military aircraft design.

Major Challenges and Problems

Working Spaces

The requirement to add 6,000 engineering branch personnel to the staff quickly, highlighted the housing shortage. Land was easily accessible. A large ready-to-erect modular, single story, air-conditioned and windowless, California style schoolhouse was purchased and erected adjacent to the main manufacturing facility. This satisfied space needs of the designers, especially those needing to produce full-size structural drawings made on large metal templates for direct transfer to the shops. A footnote to this rapid, excellent addition developed later. The supplier apparently had already sold the building to the California school system and was soon in big time legal trouble after delivery of the building to Marietta.

Technical requirements

Lockheed's major technical challenges resulted from the mistaken elimination of design margins directly relating to cruise and climb performance in the final contract discussions (and agreements) by Dan Haughton and General Terhune. Achievement of these performance targets were seriously impacted by the USAF demand for an airplane empty weight based on numbers developed by Boeing. In the final analyses, the C-5A met each technical requirement and guarantee.

Problems

The overriding concern, whether by technical, administrative, contractural, legal, or senior management personnel, had to be—and was—how to adequately cope with the flagrantly unrealistic clauses in the TPPC document. This struggle went on endlessly.

A second serious problem was meeting an empty weight guarantee that was needlessly considered by the customer as a "cast in concrete". Events would prove otherwise, but at a huge disadvantage to the program and the airplane.

New and stringent weight control measures impacted company manufacturing and material organizations. The chemical milling process was introduced, an excellent manufacturing process. However, the timing of its introduction and the procedural development training cycle added schedule pressures of an already overcrowded calendar. Subcontractors were asked to participate in this weight reduction effort. Numerous iterations of a promised delivery date resulted from this, and many unhappy circumstances resulted.

Unintended consequences.

As a result of successfully meeting this empty weight challenge the C-5A has the lightest specific weight of any large complex aircraft ever produced. In every technical area, structure, systems, wiring, paint, the C-5A Galaxy dramatically advanced the weight control state of the art.

Conclusions

The special success of the C-5A engineering program was due primarily to the large cadre of qualified, experienced technical management personnel, many of whom had worked together for many years on earlier C-130 and C-141 programs, or in the advanced design organization with the author.

"Pert-Tech" technical status tracking, reporting and control system was very effective. Coupling pert-tech results with many well structured, disciplined "trade-studies" for proper selection from available alternatives resulted in achieving design goals.

The bi-lateral approach to technical program management, coupled with the application of the USAF 375-5 series of technical management policies played a large part in the success of the C-5A technical program.

Lloyd Frisbee

Chief Development Test Engineer

Based on GELAC's data-base on military cargo aircraft design, which included records from preliminary design engineering through extended production service of the very successful C-130 and C-141 programs coupled with the continuing availability of the same design and production personnel for those programs, I felt from the outset of the C-5A competition the Lockheed-Georgia has an almost incredible advantage over all the other competitors, so far as the design and production capabilities in this type of aircraft were concerned.

However, from the day that the engineering man-hour budget for the C-5A proposal was submitted to the corporate management and rejected, a never-to-be-resolved disagreement and antipathy between engineering management and Lockheed's top management (including both Gelac's president and corporate CEO, Dan Haughton) resulted.

The projected engineering budget was substantially greater, relatively, than Gelac's previous budgets. The difference resulted, not only from the much greater size and complexity of the C-5A but also, from the unprecedented over-laying of design and cost and schedule control-documentation imposed by the largest assembly of "second-guessers" ever collected in the form of a SPO for any previous program. (As I recall, the number was in the neighborhood of 1,000 persons in the C-5A SPO).

Failure of Lockheed's top management to understand, or at least accept, engineering's explanation of the above mentioned difference between the C-5A's requirements and those of previous programs clearly pointed up a serious problem within the corporation, as follows:

The very comprehensive assessment of the C-5A engineering task had been prepared by an engineering organization with a superb record of accomplishment-the same team which had designed the C-130 and the C-141 cargo transports at Gelac.

Dan Haughton's doubt and show of no confidence in the above assessment came as a shock to Gelac's engineering organization. Adding insult to injury, this engineering budget proposal was replaced with a much lower Finance Organization estimate using the then generally accepted "conventional" method from outside sources. Gelac's president, Tom May refused to intercede.

Since the engineering design estimate of the man-hours required to design an airplane is the principal basis for determining other costs, including manufacturing costs, this unrealistically low engineering cost used in the proposal resulted in unrealistically low overall costs in Lockheed's bid.

The total lack of capable and supportive leadership at company and corporate level was conspicuous in this fiasco. This lack of leadership at the highest level reappeared again later in a situation where strong leadership might have been the means to persuade the government (DOD) to relax the empty weight specification, thereby saving many millions of dollars without any penalty in the airplane's performance. In any case, this resulted in Lockheed's famous "low bid". It won the competition but lead to the disastrous financial debacle culminating in the Lockheed fight for government backing on the 400 million dollar loan from the banking community which involved the corporation's long travail in the early '70s.

Could Lockheed have won the bid based on the more realistic cost estimates of engineering? I have always thought so. Fortunately, as might have been expected in view of Gelac's military-cargo aircraft design advantage mentioned earlier, the C-5A aircraft, itself, met all specifications but one and has established an enviable reputation in service throughout the world.

It is interesting to note that actual engineering man-hours required to complete the initial C-5A job were almost exactly the number originally estimated by Gelac engineering and rejected by Lockheed's corporate management.

Background Notes on C-5A Leaders – Lockheed and USAF SPO

Lockheed

In early 1953 I, then with Northrup, Inc. in Hawthorne, CA received a call from the Lockheed-Georgia Division with an unsolicited job offer. Dan Haughton was president of Gelac at the time. The Georgia Division up to that time was a modification center for Boeing B-47's, but had just been assigned responsibility for the Lockheed C-130 development and production program. Lockheed wanted me to put together and head up an Engineering Flight Test Development organization in preparation for the new C-130 program. I accepted the job and shortly thereafter met Dan Haughton for the first time.

Prior to this, Dan Haughton had never had an engineering organization, as such, under him, as far as I was able to learn. In all of his previous responsibilities, the operations with which he was involved were primarily manufacturing. He now found himself with a house full of engineers in a new organization, including a full-blown production-design engineering project, an engineering test organization and all the trimmings.

My previous 15 years experience had been with two aircraft companies strong in engineering and development. At Boeing, Seattle, I worked in Flight Research and Aerodynamics (principally in flight testing development) and also in the B-29 project office. Later, I worked for Northrup in Engineering Experimental Flight Test organization involved with Air Force programs such as the P-61, F-89, XB and YB Flying Wings, etc.

I mention all of the above to picture the environment to which I was accustomed; namely, companies strong in engineering, highly supportive and proud of their engineers and leaders in the field of aircraft engineering. Unfortunately, I learned at the outset that this environment did not exist at Lockheed-Georgia under Dan Haughton. It was apparent the new engineering organization was regarded as an objectionable necessity for the new aircraft program. Dan occasionally referred to it as the "high IQ boys" sardonically.

From the start at Georgia and during my entire 17 years there, this attitude did not change. Later, when I was transferred to the California Division to head up the L-1011 Engineering Program. I was in fairly close proximity to Dan Haughton as a result of the problems with the Rolls-Royce engine situation in England, which he got into personally and helped to solve. Even then, as Chairman of the Board, his attitude never changed. He continued his semi-adversarial relationship with the engineering group, even in Calac.

Dan Haughton, unfortunately, could not cope with engineers. They and their organization were an enigma to him and I firmly believe that it had much to do with Lockheed's misfortunes, all of which occurred under his chairmanship.

It is the above mentioned attitude which prevailed when as Lockheed's Chairman, he individually, personally and forcefully rejected Gelac's engineering assessment of the C-5A task. In the corporate "Steering Committee" presentation, when this occurred, not a single member of the committee, from the corporation president on down, voiced support or opposition. It was leadership to disaster.

USAF Systems Project Office (SPO) Leadership

During the period that Lockheed-Georgia was involved with the C-141 Starlifter program, Col. Guy Townsend was test director at Edwards Air Force Base where the C-141 Phase II flight testing would be done by the Air Force personnel. For some period of time, Guy Townsend had enjoyed a close working relationship with the Boeing Company. And, he had developed a friendly relationship with Boeing's Bill Cook, a highly qualified aerodynamics engineer and a Boeing executive.

Lockheed's C-141 employed a T-tail configuration. The C-141 contract required FAA certification along with the compliance with the military specifications. Because of an earlier disastrous accident suffered by a British transport with a T-tail, aircraft with this configuration immediately all became suspect and the FAA established "deep-stall"' demonstration requirements to show freedom from "pitch-up" at the stall and full control out of the stall.

The C-141 easily met the requirements. However, Dick Sliff, the FAA Flight Inspector, recognizing the C-141 was an Air Force airplane decided for "over-kill" and recommended installation of a "stick-pusher." The pusher was installed, but to my knowledge, the device was never used by the Air Force pilots. (The system has an off switch).

However Bill Cook of Boeing, looking forward to a possible competition with Lockheed (such as the upcoming C-5A competition) convinced Guy Townsend that Lockheed's T-tail C-141 has a stall pitch-up problem, based on publicity about the stick-pusher and also his knowledge of Boeing's earlier problems with the high-tailed 727. Sliff had told me that he had wished many times after he certified the 727 that he had insisted on a "stick-pusher." Cook recognized that Townsend might be a future SPO candidate and that Lockheed would probably be in the next big airplane competition with T-tail configured airplane.

> Boeing later took the trouble of building a C-141 wind-tunnel model and put it through high angle-of-attack (stall) tests to show that it pitched up. However, they were disappointed when results proved that it did not have a problem and they quietly dropped the subject. It is doubtful that Bill Cook ever bothered to tell Guy Townsend about the tests, however.

As a result of his concern, Guy Townsend tried to insist on Lockheed flight test demonstrations of the C-141 in 45 degree turning stalls. Since it was not a requirement, Lockheed refused to comply. Unfortunately, Guy Townsend may honestly have felt that there was a problem and was unhappy with Lockheed. Later, in the Air Force Phase II flight tests, his own project pilot, Capt. Doug Benefield, did put the C-141 through stalls initiated in turning flight with no problems. Townsend was fully aware of these results.

The irony of all this was that Guy Townsend was promoted and appointed C-5A SPO Director bringing his bias in favor of Boeing and his critical views of Lockheed with him to the job. (He went to work for Boeing after his USAF retirement).

At the same point in the C-5A program, when Lockheed proposed relaxation of the C-5A empty weight specification, mentioned earlier, strong unbiased support from both the SPO Director and Lockheed's chairman (which did not happen) was the kind of leadership that would have made a difference and saved the program and the country many millions of dollars.

E. A. Gustafson

Director, C-5A Wing Design Group

"Phoenix House, Southall, Middlsex, England was the home of Comprehensive Designers International, (CDI) This company was instrumental in assisting the Lockheed Georgia division during the earliest days of the C-5A design work. Several difficult problems were eliminated by the employment of engineering talent off shore."

– Eddie Gustafson

Q: What events led to the recruiting of English engineering talent?

EDDIE: There was a general shortage of in-country engineering talent. Commercial aviation was beginning a wave of purchasing new aircraft; all the major companies in the industry, Boeing, Douglas, Convair, Lockheed, Pratt-Whitney, General Electric were competing for the available people. The military, due to the buildup in SE Asia, competed for qualified personnel also.

Q: What company assisted Lockheed? What companies did you talk with?

EDDIE: Following the start of design competition for the C-5, Lockheed Georgia soon faced a shortage in finding the necessary number of engineers. Once the contract award was ours, the urgency got even greater; we had to move fast and find qualified talent in many disciplines. An agency that provided engineering personnel, CDI, advised us to consider going to England to recruit people, which is what we did.

Q: Recruiting specialized talent on a large scale seems like a huge gamble. What did you use for guidelines?

EDDIE: During the Jetstar and C-130 development days, Lockheed Georgia established a small design center in central Florida. I had a big hand in that. Later CDI provided engineers for the C-141 program.

Q: How big was the recruiting team that went to England?

EDDIE: Tom May, who was C-5 program manager at Marietta at the time, sent Whit Holland, me and Everett Lampkin, an administrator, to make the first contacts and evaluate the possibilities. Three days later I called Tom May and stated that it would work. May responded with the statement, "OK, its all yours, get it done". During the next two weeks we leased a building in Southall and started putting the workplace in order. At that time, we leased the five floors of the seven story CDI building; it was walls, floor, and a roof. We needed to do just about everything: install proper lighting, and provide desks, engineering drafting tables and all the associated equipment. We bought drawing boards from Germany, and drafting machines from Italy. We equipped the building with partitions from a contractor in Liverpool. By the way, the desk suppliers were one of the first to note a penalty clause for non-performance in the contract. This was a new item in the way of doing business for them, but they agreed. They set up assembly on one floor of our quarters to make scheduled deliveries. One of the products in the engineering design process is the blue line drawing. We needed a large "Blue line" copy machine. It took some fast talking to get it in our facility in a timely fashion. Blue line machines were like many necessary engineering tools in short supply. For instance, we issued a memo to our Marietta counterparts: when you visit England bring a drafting machine with you in your luggage(this was years before CADAM and FAX). After the contract go ahead I returned to Marietta for about three weeks to turnover my C-5 fuselage design duties and get procedures set for communicating with the wing design people from the Lockheed design group located in London, really in the shadow of Heathrow airport.

Q: English engineers must have had some of their own ideas on how to operate. Were there any serious confrontations?

EDDIE: None. By the time I returned Everett Lampkin had already set up a two week engineering procedures course to cover details of Lockheed engineering methods. This orientation was very successful; we used it for indoctrinating all the people who would affect any design work. However, we were really getting some especially skilled people. British engineers of that time had trained formally in college while working in factory shops rigging, welding, or other manufacturing tasks, to round out their skills.

Q: What made these valuable personnel so quickly available? Didn't you get some flack from the aviation industry moguls in England?

EDDIE: The aircraft industry in England had just experienced the cancellation of the TSR-2 fighter program. Many of the engineers were working odd jobs, not directly in any engineering capacity. One important point was that the people did not wish to emigrate, or even work temporarily overseas. Also, the English government was very vociferously opposed to the potential "brain drain" that could occur. I had an interesting experience that took a couple weeks to straighten out. My passport was lifted by the Home Office for three weeks when it was learned that I didn't have a work permit. Work permits have to be applied for by someone working with the company in-country; until I moved over to England there was no one to request work permits. In the meanwhile, a bill was introduced into the house of commons to have me deported. It was subsequently killed by my M.P. friends, who understood the full scope of our intensions.

Also, we early on developed an agreement with all the English firms that Lockheed would not make offers to anyone whose home employer did not wish to lose. We then placed work with Hawker Siddley, BAC, and Marshall of Cambridge and developed good working relations. The total number of people working on the London C-5 wing design project at its peak was 985 in all. The ratio was 10 percent Lockheed from Marietta and 90 percent from England. The design work was divided into discrete packages that addressed the inner and outer wing sections. Design work on the center wing was completed in Marietta.

Q: Getting and sending information must have been a problem initially, was it?

EDDIE: We leased a phone line from the telephone company that gave us 24 hour a day, exclusive use of that line/cable at a cost of $1800/ month. We set up a routine that had communication 20 hours a day; of this, six hours a day was devoted to computer use. During the day we used voice, and teletype during the night. The Marietta computers dedicated to stress computations were run by engineers in the London design team. This scheme was an evolution that paid huge dividends. The cable broke at one point during our lease, but the telephone company was right on top of the problem, finding the cable break within days, fishing up both ends and making repairs within a week. The other method of data exchange was to send completed drawings in tubes, via USAF APO back to the plant in Marietta. We could get two day exchange of data most times. This English branch had to contend with pretty significant changes that the Marietta group dictated. A big change to improve wing lift was changing from the initial hinged trailing edge flaps to the Fowler flaps. This change was made after we were well into the production design of hinged flaps. Also requiring leading edge slats to be sealed around the engine pylons added rework of the basic designs. I believe this is the only aircraft with slats sealed on both sides of each engine pylon. In the early going, the leading edge slats were slotted, then sealed. We looked at sheet metal construction, and honeycomb possibilities, and finally back to sheet-metal. The early design work showed that the wing needed more structural weight, and this data was relayed to the USAF. Their response was to have GE develop an engine with higher thrust, which Lockheed would have to sponsor and pay for added development costs. This could not be accomplished even if schedule relief were possible, which was never even considered. This design team predicted wing fatigue failure long before the failure was demonstrated on the ground test fatigue article. This data was transmitted to the USAF before all engineering drawings were released.

Q: What number of hours per week were you averaging to keep on schedule?

EDDIE: We had a very enthusiastic group. Yes, they worked overtime, and were paid for it, though little by US standards. I also found that there were fellows who came in well after hours to continue working on details on their own time, primarily because they knew that this bird was something special, and it's going to fly. This was done without extra pay

Q: Did this design team stay together, and did any eventually come to the States to work?

EDDIE: The group was together for a period approaching 22 months, and as the workload began to wind down I went to companies like Boeing, Fairchild, Douglas and CALAC to get additional work, with moderate success. The competition for the SST was in being at the time. Lockheed Corporation, in their proposal, stated that the British design team would design a major portion of their SST.

Q: This project must have had a number of non-scheduled, but interesting, surprises.

EDDIE: It certainly did. This outfit had its own flying club, its own bowling league, and two very wonderful winter holiday seasons. At Christmas I took the opportunity to go around to each of the groups to express my appreciation for all the fine dedicated efforts. Since I was a managing director, this sent out mild gossip ripples, since the management-worker divisions were still pretty strong at that time in the U.K.

Also, Lockheed hired a PR company to rebuff the tabloid storytelling that was inevitable in the beginning of our efforts. That resulted in a considerable easing of tensions on each side. There were the series of visitors from the plant, the USAF, and the local political scene. I was always hoping some of the royalty might make a visit, but no such luck.

There was one rather nice event that really displayed the cooperative side of the local police. I was visiting in the US and had to contact tool design personnel in the Potters Bar area. I knew that men were working there, but it was late on a Saturday afternoon and the switchboard was off. So I called the local police station, and after a few minutes of explanation I was assured that one of the Bobbies on duty would bicycle on down to the shop directly. Within the next thirty minutes I was in contact with the shop in Potters Bar. Then, there was the flap about our storage vault for our drawings. We built a room in the basement area, then lined it with concrete block to assure a fire-proof quality. We had to convince the authorities that no one would work in that room, even momentarily; otherwise we would be required to build a window in the vault, which would have defeated our fire proofing and security plans.

Q: In the final tally how much did the budget suffer?

EDDIE: The reason Lockheed went to England is because the engineering talent was there. Cost was a consideration, but was not overriding. The all up costs for salaries, travel, leases and renovations to the work areas was nine million dollars. If engineers had been available in the US the cost of salaries would have brought the cost of the design effort to double that figure.

Robert B. (Bob) Ormsby

Deputy to the C-5A Engineering Program Manager 1965-1967

Q: What was the principal work of the ops research group when you joined engineering?

BOB: Operations Research was an outgrowth of World War II experience where it was found that many existing weapons systems were not being used to full effectiveness. In the 1950s the Air Force initiated efforts to develop a nuclear powered strategic bomber. This included major research into the associated nuclear propulsion systems. Lockheed, General Dynamics, General Electric, and Pratt Whitney were involved. This engineering work also led to conceptual studies of a nuclear powered transport. By their very nature these airplanes were very large, grossing 500,000 pounds and upward.

Q: What in these studies were you able to carry forward to the CX-HLS study which resulted in the C-5 operational requirement?

BOB: Computer routines that were written took into account pertinent variables of the airplane, including the cargo compartment dimensions. Every piece of Army equipment that might be airlifted was in the mix. The studies then ran mission sorties, delivering total requirements over a given distance on a fixed schedule. This computer study also addressed the proposed airplane gross weight limits and range limits. Among the results were numbers that sized a cargo compartment that was efficient for military cargo. The nuclear powered aircraft studies ended in the early 60s but they made a serious contribution to the information bank for airlift of heavy, oversize military cargo. Later, the CX-HLS studies were guided for SecDef MacNamara by John Keller. He incorporated results of these analyses in the Heavy Lift requirements.

Q: Leo Sullivan, Chief Engineering test pilot at the time, specifically rejected ejection seats for the crew, because in his view, he felt that nuclear power required special procedures in every area, especially flight crew responsibilitiy. Were you in on those discussions?

BOB: No, I was not. However, I vaguely recall some conversation along those lines pertaining to the WS-125A, (the nuclear powered bomber study). They may have indirectly applied to the cargo airplane study as well.

Q: What were your responsibilities after the company won the C-5 contract?

BOB: My assignment was deputy to the Engineering Program Manager, Gibby Gibson. My engineering developmental role was to oversee the aerodynamic performance of the airplane. All that ops analysis and wind tunnel experience was invaluable here.

Q: In the development process did aerodynamics concepts override demands of other disciplines, such as structures or engines?

BOB: That is not the way the system works. There is constant interaction of requirements for each design discipline. As development progressed, the gross weight of the C-5 increased from 680,000 to 700,000 then 712,000 pounds. At this weight the C-5 would barely make its performance guarantees. The engine max thrust output was 40,000 pounds. When the designers found that added weight was required, the company turned to G.E.

Q: What kind of a reception did that event trigger?

BOB: Well, they smiled and said, "We were expecting you". The result was we came to an agreement. G.E. could provide the added increment of thrust, at a cost to the company of $5 million dollars. Lockheed agreed to foot the bill, but the SPO put a terrible clinker in the process. The SPO insisted that if the empty weight of the airplane exceeded the guarantee figure, the USAF would never accept a single airplane. The immediate question put to the customer was: "What difference does a couple thousand pounds increase in empty weight make if the airplane still meets its takeoff, landing and cruise performance requirements with the added thrust?" No very intelligent answer was forthcoming, then, or at any later date. And, indeed, there wasn't any common sense to that position.

Q: Was that the last resort?

BOB: No, we did get to put the case to the Assistant Secretary of the Air Force for R&D, Al Flax. He listened patiently, but sided with the SPO. Thus, Lockheed was forced into achieving exceptional aircraft performance along with putting together an incredibile weight saving effort.

Q: Were there any previous or current design efforts that could be cited to help the company make its case?

BOB: Yes, but the inflexible attitude that existed in some DOD circles made any change impossible. One can examine the comparison of weight growth histories between the C-5 and the B-747 to get a perspective of the problem. Boeing was in a sales effort to market their B-747 to the commercial airlines. I had access to data showing the B-747 gross weight growth plotted against months from go ahead. If the same data for the C-5 were plotted on the same graph, the gross weights never deviated by more than two or three thousand pounds at the same point in time. Like Lockheed, Boeing ran into a performance limitation when the gross take-off weight of the B-747 exceeded the weight of about 712,000 pounds. Boeing's airline customers couldn't care less if the weight increased by a few thousand pounds, *as long as design payloads could be flown guaranteed distances while operating from specified fields.* This required that Pratt Whitney, the engine supplier for the B-747, be able to increase the thrust output of the JT-9D engines. This was accomplished, and the flexibility of this approach was a keystone of the success of that airplane.

Q: Didn't the airlines also develop a technique called "balanced field" take-offs?

BOB: Yes, this was a clever method taking advantage of the conditions existing at the time of each individual take-off. Operationally, the gross weight of the airplane is seldom at its design limit.The passenger load factor may be down, the field length may be well in excess of requirements. All variable factors are charted in performance data that govern flight procedures. Thus, lower power settings can be safely used for existing conditions. This flexibility extends engine life. The Air Force did recognize the pluses of this method and adopted it.

Q: The "brick wall" limit for the airplane empty weight was met. What were the company solutions to solve this challenge?

BOB: Substantial aerodynamic refinements had to be made. It was my job to oversee this intense aerodynamic design refinement process. It was my good fortune to work with the best aerodynamic specialists in the world. Frank Wilson, Bill Johnston, Jack Paterson, Dallas Ryle are just a few names that come to mind. They led a group of engineers that "sweat bullets" to achieve the exceptional takeoff, landing and cruise performance targets Of course, this was done while achieving outstanding levels of stability and control. General Electric succeeded in delivering an engine that produced the specified thrust and with the specified fuel consumption which made it all possible—after the aerodynamic refinements were perfected and included—to achieve the contract cruise peformance. Extensive wind tunnel testing preceeded the selection of the wing airfoil section. Another difficult challenge was assessing the scale effects, or Reynolds number difference between the wind tunnel test conditions and full-scale free flight conditions of this very large airplane. Wing to fuselage fairings and landing gear pod fairings had to be designed to eliminate air flow separation. No external area of the airplane was exempt from very detailed analyses for possible drag reduction. As difficult as it was to meet the cruise performance requirement, making the takeoff and landing requirements were, no doubt, the most difficult. With engine thrust fixed and empty weight growing, it became necessary to increase lift in the flaps– extended configurations, but with no drag increase.

Bumping the engine thrust limit, while meeting exacting take-off, landing and cruise performance targets, also resulted in the inauguration of a very detailed, all encompassing weight elimination program. The success of these efforts is in the books. The C-5A met or bettered every performance guarantee in the contract.

Q: From the perspective of 33 years later, what is your comment about the engine?

BOB: The GE engine performed in an outstanding manner. In this case that adjective is well deserved. This engine was not an off-the-shelf product; engine development paralleled airplane development, thus a number of mechanical "growing pains" were expected. Due to G.E.'s expertise, those problems were very minor. In the critical CAT I and II flight test programs, unintentional in-flight engine shutdowns were nil. In this case, GE never received the acclaim they deserved.

Ed Shockley

Chief Development Test Engineer, Lockheed Georgia 1964-1969

Q: Ed, how did you chose aviation, and what were the circumstances surrounding your joining Lockheed?

ED: In March 1950 graduation from Georgia Tech allowed me to take a job with Douglas Aircraft. Having been a US Navy pilot during the war I applied for flight test; but their chief pilot, Johnny Martin, didn't want engineers as pilots—eliminates arguments he said—thus I worked as a flight test engineer. Lockheed Georgia division hired me as a flight test engineer in 1953, that's how I got on board. My assignment was at the Burbank facility during construction of the two YC-130 prototypes. Flew as copilot on Connies and the YC-130's during my Burbank tour. I wrote the first pilot's handbook for the YCs, a forty pager; no, it didn't make the best seller list. When the C-130 moved to Marietta I volunteered to come with it, stayed in flight test division, moving up in management during the C-130, Jetstar and C-141 development cycles. In 1964 my job as Chief Development Test Engineer gave me the task of developing the C-5 flight and ground test programs. Thus, during the early C-5 years my view was on the test area, though a bit remote from the actual flying. However, I had a close view of the many static tests that were part of the ground test verification programs.

Q: Wasn't monitoring static tests kind of boring?

ED: That's a very uneducated point of view! I happened to be on hand when one of the tests involved actual wing failure, a truly depressing event in that highly charged atmosphere. But, to everyone's credit, that problem was taken in hand properly. Airplanes in flight test had an 80% weight limit already in place. This is an industry standard practice until ground tests are complete, so the immediate focus became the a wing fix. This wing design and implementation followed in rapid order. The problem was relatively minor to repair.

Q: In your position did you deal directly with the SPO, Gen Townsend, how did that go?

ED: Yes, I did. At the outset one point must be understood, the SPO was the last level of discussion for problems that could not be resolved at any intermediate level. And the toughest, longest discussed, most thoroughly analyzed condition was that of C-5 empty weight. The number was one of the original contract guarantee points, and as such had a severe penalty attached to it in the infamous TPPC. Townsend was viewed as a Boeing man, that is, he made his reputation as flight test director for the B-52 at Edwards AFB. He made no secret that he thought Boeing should have been selected the C-5 program winner. We had several interesting discussions on the weight subject; but he would not move off the "magic number".

Q: Can you give a specific example or two?

ED: Certainly, there was a USAF wide thought, procurement side as well as the development, that this, the world's biggest airplane, would have plenty of capacity to be "all things to all people." The realism of the requirements was lost in this pervasive idea. Exposing the airplane to an environment where a foot soldier could bring it down with a shoulder mounted missile was faulty thinking. But such issues as providing a terrain following capability to fly a track at thousand feet above ground level stayed in the book for a long time. Another requirement that Lockheed felt was overkill was the crosswind landing gear(CWS). We designed it, installed it, made it work; but also pointed out to our customer that the weight penalty and gear complexity was more than common sense required. The airplane can be routinely landed in a crosswind using a crab technique, realign just prior to touchdown and complete the landing. To my knowledge, the CWS was very little used. The weight saving on this CWS would have alleviated the extended weight reduuction efforts that were practiced in the manufacturing cycle — chemical milling procedures; reducing the number of internal access positions on the wing and fuselage, (these reduce weight, but increase the maintainability factors). Lockheed felt that because of this rigid mindset that the USAF was not getting the best possible airplane. Townsend and I crossed swords on this and other issues several times. That airplane empty weight number didn't change.

Q: Did the vendors get involved?

ED: Our design people worked with everyone to fight the weight penalty, some were cooperative, others decided to leave the program entirely. Each one of these events piles up the man-hours and cost. The impact of the large number of requirements cited for the program was huge. We(Lockheed) found numerous occasions when direct contradictions had to be resolved at the SPO level; this procedure was very time consuming. I have to add this note to this tale of woeful recollection; years later at the annual SETP meeting in Los Angeles the corporation sponsored a hospitality suite. General Townsend visited while I was there with my wife Dot. He joined our circle and says to me, "despite all that has transpired, I want to say that the C-5 Galaxy is the best aircraft the USAF has ever bought." I didn't get physical, tried to smile and left it at that.

Q: In your senior position you must have had more than a little contact with the media, especially when development issues arose that got in print with only partial facts. How does this type of negative information get straightened out?

ED: Lockheed's publicity department was the first line of contact most times, we were briefed, however, on the techniques in current use by video reporters which stressed the "sound bite" technique, so rigorous reporting of technical data was seldom the case. Roller and Frank, a public relations company, was engaged to give management a short course in how to cope with these reporters. I sat through a couple hours of it and I learned that they championed Herman Talmadge's technique, to whit, answer the question seeking sensationalism this way. "I'm glad you asked that question," then tell them what you want them to hear.

Q: Well Ed, how long did it take before you had an opportunity to get aboard a test airplane, and did you get a chance to fly?

ED: Believe it or not, it was actually about three to four months before that opportunity was present. You were there, and know the pressure there was to get data and make the program productive. However, I did have my chance, it was either Walt Hensleigh or Glen Gray that let me handle the controls while at altitude.

Q: Here's your chance to be nonpartisan in your comments, is that possible?

ED: Certainly, it was an impressive experience; because, despite reading previous evaluation statements by Walt or Glenn and listening to customer comments by Jesse Jacobs and his group of pilots you just do not get the impact of the excellent control response this huge airplane has until you fly the bird yourself. I really felt like, if I could take a sabbatical for two weeks I'd press on Walt (the chief pilot at the time) to let me fly. I'd do it for free.

Q: Later, you were deputy C-5 program manager, did SPO's meetings have much change?"

ED: The USAF officer in the SPO at that time was General Newby. We alternated monthly meetings between Dayton AFB and Marietta. Requirements consolidation, elemination of duplication was a constant thorn. For example, I got him to understand why a $35 hot cup which the USAF could get at any restaurant supply store cost them ten times that number. We detailed their contractual documentation requiring us to provide an illustrated parts manual, a maintenance manual, periodic updates, fireproofing, crash test to nine "G," for said hot cup as the clinkers in the process. He vowed to get that stuff changed. Seven months later he admitted that the process was institutionalized. We deleted that item from our monthly agenda.

Ric Johnstone

Director, Flight Operations, L.M.A.S.

RIC: My Air Force pilot training began in early 1970; my initial squadron assignment was with a C-130 air rescue squadron in the Philippines. About mid year 1973 orders came through to report to Dover AFB for training with the second group of pilots checking out in the C-5A. We finished our training near the end of 1973, just in time for me to get involved in a few flights to Lod airport, Israel on the "Nickel Grass" operation.

Q: What were the transitional problems for you, or did you have any at all?

RIC: Nothing of very long duration. Actually, the C-5 airplane, for me, has always been easier to fly than the C-130. The Galaxy is a lot more stable in every configuration. But where a pilot notices it most is during an approach to landing. Where the C-130 tends to wallow a bit, not too easily trimmed in the lateral axes, the C-5 is just the opposite, solid as a rock. Obviously the designers had this intent, and they succeeded quite well. Then, the problem of flaring on landing does require a goodly amount of rethinking. Your eyes tell you that you are 60 to 70 feet in the air, but the radar altimeter is reading zero as the aft trucks slip on the runway, or bash it, depending how well you have managed the flare. Another area that gets the new pilot's attention early, is the ground maneuvering the C-5. One feels very conscious of clearing the wing tips, and where the engine thrust is being directed during the necessary turning, maneuvering out of parking, into narrow taxiways etc. The rear gear castering function really is a key to the ease of ground maneuvering.

Q: What about the height of the cockpit? Was this a big transition item?

RIC: Well, not to any great extent, probably because the visibility out of the cockpit is excellent. The height does give a different visual clue for estimating ground taxi speed; once again, just something else to recalibrate in one's mental checklist.

Q: Did the cockpit instrumentation look strange initially, after all, there are no instruments in the Hercules with flight or engine data presented in taped form?

RIC: I've always been a fan of having flight instrument information in the taped format with a digital backup. For me, this cockpit instrument panel layout was a big step forward. I find it easier to read the numbers and felt the new instrument arrangement contributed to greater accuracy. resulting in getting the job done more efficiently.

Q: What are some strong impressions you had from flying in "Operation Nickelgrass"?

RIC: We flew missions that airlifted concentrated high gross weights of ammunition, then on a couple flights we did carry trucks, armored personnel carriers and the like. These missions were slightly over 730,000 pounds. At the outset the airplane had been operating under a gross weight limit of 710,000 pounds. MAC issued a waiver for the duration of "Nickelgrass". Then we had the old problem of needing to use Lages in the Azores. We also had in-flight limitations on whose airspace could be used; thus the enroute time was greater than necessary. It was during "Nickelgrass" that IFR became a high priority.

Q: When did you get into air refueling training?

RIC: Operation Nickelgrass was completed long before any extensive IFR crew training and re-qualifying could be scheduled. The training was accomplished using a KC-135 as the tanker. The pilots that I flew these missions with were Dick Rammage, Josh Hinson and Mark Thomas. We were the first pilots to get re-qualified.

Q: IFR is a special brand of flying, sucking fuel out of an airplane half your size. Did you find this experience particularly trying?

RIC: Nope. Not difficult, nor easy; just a business worth getting done.

Q: Daylight and night training?

RIC: Yes

Q: What followed then?

RIC: I did line flying through 1976, got off active duty at that time, which coincided with the general cutback of the armed forces, to concentrate on my business career. I went to work for Lockheed Georgia in 1979 in the marketing department. I transferred to the flight operations department in mid 1985. When President Reagan began the strengthening the military in the early 80s, I joined an Air Force Reserve squadron in Dover, Delaware flying C-5s. And, I've been with the Reserves ever since.

Q: Does that mean you had a chance to get involved in Desert Shield/Desert Storm?

RIC: Yes, certainly it did. We flew quite frequently in the early months of 1990 with the hope to delay the need for the call-up, because it did impact our civilian jobs pretty severely. However, the official call-up came on 1 September 1990. Ten months later we were returned to inactive status. War material was airlifted from places like Hunter Field, Savannah Ga.; Pope AFB in North Carolina; then flown overseas. The bases in Europe that experienced a heavy amount of activity during this period were Ramstein, Germany; Mildenhall, England and Lages in the Azores. I recall one flight approaching Ramstein counting twenty three C-5 and C-141 airplanes on the tarmac at one time. We were most always near gross flying to Saudia for the first several months of the buildup. Initially, the equipment was all types of rolling stock: trucks, armored personnel carriers, Humvees, ambulances, tanks, you name it. Also, the early flights out of Saudia were light loads: equipment not repairable in country, emergency leave personnel and similar cases. The C-5 flew very well during this high demand period. Yes, we did fly with some minor problems, like an inoperative roll axis servo in the autopilot, or something that had no impact on safety. The mindset was "do not restrict the mission by stacking a series of improbable what-ifs." For the eight hundred and sixty hours (860) that I accumulated in that period I had no in-flight emergencies of any kind, no engine shutdowns, no gear problems, and no aborts.

Q: What form of crew flight limitations were required?

RIC: The flight hours of 125 for 30 days, 250 for 60 days and 330 for 90 days soon became 150 for the first 30, 275 for the first 60 but, 330 for the 90 days was ironclad. We started off flying with augmented crews; three pilots, two flight engineers (FE), and three loadmasters (LM). A third LM is required by regulations when carrying more than 33 passengers, and eastbound we always did. The C-5 has 75 seats in the upper deck, and they were always filled. After about four months the LM crew requirement was reduced by one, and we operated with only two. The theory was that the loadmaster could accumulate some crew rest airborne on an eight or nine hour flight. Practically, it became a very difficult workload in most instances. There were occasions where some crews were flying into Saudia, back to Germany then back to Saudia, over a five or six week period without any opportunity to get back to the States. Others would get one trip into Saudia then be scheduled directly back stateside. These crew scheduling inconsistencies did not occur too often, however. Weather never was a real problem. The minor delays that did take place were in icing weather that occurred in Ramstein or Frankfort, Germany. Ground deicing of several C-5 fuselages and a half dozen C-141s soon uses up all available deicing fluid.

Q: Having been through a concentrated flight period like the Desert Storm operation and with a long term Instructor Pilot experience, what are your highs and lows in the airplane?

RIC: At this point I have over 4500 hours in the C-5 Galaxy. I have only had to shut down one engine for cause in all that time. That was a hydraulic leak problem. Once, on a training flight during a touch and go series, the right forward bogie canted off twenty degrees to the right, probably a malfunction in the cross-wind gear circuit. Airborne, aware of our problem, we knelt both forward bogies and made a full stop landing on the nose and aft mains only (made like a C-141). The handbook procedure was followed for this problem. What I really feel is little used by the Air Force is the short field capability of the airplane. Everyone who has any experience in the airplane knows this airplane can be stopped in very short ground rollouts, because of its outstanding braking system. After all, there are twenty four wheels on the four trucks, each with its individual brakes. In the late 80s combat low level approaches were inserted into the training cycle. This flight technique involved spiral letdowns over the field of intended landing–to avoid perimeter small arms fire– with a spot touchdown 500 feet from the approach end of the runway followed by a very short roll out using the normal braking system. Crew confidence in the airplane soared. It was like recognizing a gold nugget in the taxiway that you had bypassed for months. But Desert Storm brought on different requirements. Long runways at every stop negated spot touchdowns. The special operations crews, probably, are the only pilots competent in the short field capabilities of the C-5.

Glenn M. Gray

Engineering Test Pilot, C-5A Performance Testing

Q: You were one of the first five pilots to fly the C-5A. Did Leo and Walt have a very extensive checkout process?

GLENN: As you know, we grew up with the design, practically on a daily basis for at least 18 months before getting our turn at the yoke, so there wasn't a lot of discussion necessary. What impressed me from the outset was the flight control improvement over the C-141, not that the Starlifter was so bad. The Galaxy was just much better, especially in the approach configuration.

Q: What flight testing occurs prior to performance series?

GLENN: After the flights on ship 6001 proved it airworthy, a series of stall tests were flown to get data on every flight configuration pertinent to the operation of the airplane. Certainly other primary equipment was being evaluated, but the emphasis was on the airspeeds for the low side of the flight envelope. As ship 6002 was the second C-5 off the production line, –remember this was a concurrent production and test contract, –we had to squeeze everything we could out every flight. Preliminary climb performance tests were also flown. All these flights were accomplished in the Marietta, Georgia test areas. However, ship 6002 (and crew) spent the majority of its eight-month test cycle in the friendly confines of Edwards AFB in California. Lockheed had most of us living in "double-wides" south of the base.

Q: What specifically is flown in the performance test series?

GLENN: Using the stall speed data, takeoff, climb, approach and landing speeds are computed. Then you fly these test points to demonstrate the validity of the estimated performance of the airplane. There is a host of related testing that must be done. For instance, recommended rotation speeds for takeoff are certified, then we must show what happens to the airplane when these speeds are abused on the high or the low side. The low speed demos get the airplane into a very high pitch attitude before it leaves the ground. The flight test planners had a wooden skid installed over the skin on the tail of the C-5 as a protective measure. It really wasn't necessary. The wooden skid was several inches thick. We did mark the runway on one of the min–speed early rotations: had the wooden skid not been there our tail clearance would have a "sufficient" by that couple of inches. The matrix of tests points include all pertinent gross weights. Takeoffs and landings using the three different flap settings are required to verify the forecast field distances. Climb gradients with four or three engines operative are also flown to acquire data. All this number gathering gets a bit monotonous, except for those aborted take-off tests. Demonstrating aborted takeoffs at the 728,000 pound weight was the real revelation. The technique used is to accelerate on four engines to the test speed, 151

Radome off flight test at Edwards AFB.

knots as I recall, fail the critical engine (chop the fuel), then bring the airplane to a stop as rapidly as possible. Here we learned how effective the braking system is on this airplane. The brakes do a superb job. We only needed to use about two thirds of the 12,000 foot runway, always stopped with plenty of runway margin. Those aborted take-off tests provided a high point in excitement.

Q: Were the effects on the airplane? Anything unusual in these aborted takeoff tests?

GLENN: This airplane was equipped with a water spray system for brake cooling. The test requirement demands that the airplane remain untouched for a full five minutes after stopping, to demonstrate that no added system failures result from the stress of the hard deceleration. As you are aware, the main wheels of the C-5 have metal plugs that melt at extreme temperatures, releasing tire air pressure to the atmosphere. This safety feature prevents tires from exploding off the wheels resulting in debris flying through the wheel well and surrounding area. We did have several flat tires in the five minute wait period, but we got through the test demos OK. So, to answer the question, we did develop several flat tires. We also used the water cooling system on occasion; however, it was also possible to get airborne and use air cooling of the wheels and brakes with very good effect.

Q: Did any USAF pilots get to fly any of this spectacular performance?

GLENN: Although the company had a full crew on the airplane, Frank Hadden and Jerry Edwards were in my crew, we did have Air Force participation in the seat from time to time. Jesse Jacobs and an RAF exchange pilot on duty at Edwards, John Miller, got quite a bit of exposure. By the luck of several neat coincidences, our flight test engineer was Allen Youngs, a Brit to the core. He made the conversation lively for Major Miller.

Q: Did you use reverse thrust to stop the airplane on these simulated aborts?

GLENN: With one engine out, only the inboards were reversed. The wheel brake system does 95% of the stopping of the airplane. With anti-skid installed, the pilot can literally stand on the brakes to complete the stop

Q: During the early flight testing, you had to adopt a different technique for setting maximum power on the engine; would you explain this?

GLENN; For best performance, shortest field length for takeoff, one needs to have the maximum power set at the instant of brake release. This engine had some unusual, predictable characteristics. At max power, with the throttle set and fixed (friction lock on), the engine pressure ratio would droop after 15 to 30 seconds. We would spend upwards of two minutes stabilizing power before releasing brakes. The heat generated by the engine dynamics at this high power caused most of the reaction we read on the engine instruments. The airplane was positioned one thousand feet down the runway in the direction of takeoff, stopped, brakes set, maximum power set, cook the engine for 90 to 180 seconds, release brakes and complete the takeoff. These days, since the airplane and its performance is well documented, the USAF has altered the takeoff proceedure to suit the operational needs.

One of these engine power setting events had a disturbing, yet fortunate, development. The particular test in progress required that the number one engine be shut down at airplane rotation and continue the takeoff. While we were on the runway completing the 90 second engine "cook" procedure, the number two engine fire warning light illuminated. Naturally, this requires rapid power reduction and engine shutdown. Had this event happened two minutes later there would have been an opportunity to fly an unscheduled two engine climb-out profile, and at that weight we didn't have any desire for that. Maintenance personnel later found that one of the burner cans in the number two engine had a hole burned through it. Bad news had a timely arrival that day!!

Q: Was the Stablity Augmentation System (SAS) installed from the start of testing?

GLENN: The SAS was not available for the first seven to eight months of testing. We managed quite well without it for the take-off and landing performance tests. The system contributes to flight efficiency in cruise, and especially in the IFR task. We did do some qualifying refueling flights behind a KC-135 tanker. For these tests we had a very experienced B-52 pilot with many hours of IFR ops in his log. He quickly let us know the C-5 flight control made the task quite routine. These tests led to some interesting test results and humorous recommendations.

Q: Oh, what was that?

GLENN: Well, you know that the C-5 is twice the size of the KC-135 tanker, and in 1969 they were operating with the original low thrust engines. Their gross weight takeoff performance was less than spectacular on hot days. Our testing required a couple dozen hookups be completed to wring out all the fine points of flight technique. The other heavy requirement had the tanker delivering three hundred thousand pounds

(300,000) of fuel to the C-5. Well, that contract item would have required the tanker go back to base, reload, takeoff and rendezvous again, at least three times, to get that square filled in on the test form. However, because both the C-5 and the tanker could reposition their respective transfer valves to reverse the fuel flow, allowing the receiver to refuel the tanker, we first took on 60 or 70, 000 pounds, then gave it right back to them. Set the valves for normal transfer again. Four or five of these cycles got the job done quick time. After the tests were over, in the post flight debreifing, we reminded the tanker crews that with a C-5 around they no longer had to make heavy weight takeoffs on hot days. Just have a partial fuel load, hook up with a C-5, top off the partial tanks, then get on with their mission. (The KC-135 tankers were later equipped with higher thrust turbofan engines.)

Q: What airspeed and altitude did these IFR tests use?

GLENN: We worked in a block altitude of 20,000 to 24,000 feet. The speed was 250 knots for this preliminary series. The only noise was the knock from the tanker probe plugging into the C-5 refueling receptacle. After that it was a close order flight drill with all eyes forward and up, except Jerry's, of course. He was busy making certain the fuel filled the proper tanks in each wing.

Q: What other test flights did ship 6002 have assigned to its schedule?

GLENN: The US Air Force wanted to know what problems to expect if the nose radome was damaged or completely torn off the front end of the plane. So we had a test flight that flew around South Base with the sixteen-foot nose radome removed before flight. Our airplane was equipped with a nose boom Thus, we had both ship's production pitot static and instrumentation pitot source to give us altitude/speed references. Both systems worked perfectly. The air noise was extremely loud, and we kept the speed below 200 knots. It was one of the shortest and noisiest tests I've ever flown.

Q: Did you have to fly any cross-wind gear take-offs or landings?

GLENN: Yes, we did several. It's a very unusual feeling sitting in a position that is at an angle to the takeoff direction. We did a fairly long preflight brief before the first one, as none of us had any recent experience in any large airplane-I didn't have any, before or since. The pilot has to look out his clear vision window, his head is turned full left. Only the first takeoff and landing promoted a bunch of laughs. After that it was all drill. These maneuvers were performed on the lakebed runways. The wind was fairly steady, but occasionally gusted up to 45 - 50 knots. We only did the maneuvers with a right cross wind; this allows the pilot look out his own clear vision window, thank you. I suppose one could say that these maneuvers constituted some fun flying.

Til Harp

C-5A Pilot

Q: You have a C-5 flight experience that is truly unique, would you share a few details?

TILL: You are referring to the C-5A crash that occurred on 4 April 1975 in Saigon, Vietnam. Yes, there are details worth relating. Although there were tragic results, the efforts of the survivors were heroic, and are worthy of note. Our assignment required us to fly a cargo of howitzers from Clark Airbase in the Philippines to Saigon, a three hour flight; then airlift as many personnel as possible back to Clark. We had fifteen flight personnel and a medical crew of four nurses who normally would staff an air-evac C-9 aircraft. Due to the circumstances existing at the time, that is, minimum friendly control in the immediate area, (Saigon was surrounded by the North Vietnamese) we had a blanket clearance from the highest level of authority to evacuate as many people as the airplane could reasonably carry. After landing and parking the C-5, we were rapidly surrounded by numerous buses filled with people from the orphanages with dozens of babies and small kids under the age of seven looking for a way out of Saigon. We had to get the terrified adults to back off until we had a chance to unload the armor we brought into Siagon. We had expected some problems, but this set of circumstances was bizzare by any measure. Our ground time was about three hours. In the loading process we put close to (200) two hundred kids, babies and responsible adults, in the troop compartment. Arm rests were removed and six to eight kids were strapped in each three seat row. Securing people in the cargo compartment was accomplished with tie down straps for "make do" seat belts attached to the cargo floor. Our loadmasters and flight engineers boarded another 180 to 190 people in the cargo compartment. The loading drill was very difficult under the circumstances. There wasn't an accurate head count; rescue was the thought that stuck in everyone's mind.

Our departure time was 4 PM local time. The climb-out was standard, 250 knots through 10,000 feet, accelerating to 270 knots as we continued the climb to cruise altitude. We had just passsed the Vung Tao radio beacon, which positioned us about 30 miles out over the water when our problems got much worse. The aft pressure door, which is 13 feet by 19 feet in size, swung out of its vertical, flight position as we climbed through 23,300 feet. It swung violently upward, smashing into the hydraulic lines and the flight control cables in the vertical tail. Within seconds, the effects of rapid decompression were felt in the cockpit, a fairly strong white mist forming everywhere. I flipped on the seat belt light, called for the appropriate checklist and put my oxygen mask on. The pilot made an immediate decision to return to Saigon airbase, declared his intent and began a turn back to base. While this minimum maneuvering was initiated we were able to determine that troop compartment personnel were in reasonable shape and the crew members were on oxygen. However, the loadmaster in the cargo compartment reported that the pressure door was missing and several injuries had occurred there. Within less than a minute of the pressure door loss, our normal flight controls, elevator, rudder, and half the roll control (the left aileron) were immobile. Hydraulic systems one, two and three had lost pressure. This also immobilized the pitch trim as well. We had a couple dozen alert lights illuminated on the pilots "caution panel" as well as lights overhead on the hydraulics panel. We learned later that the pressure door motion into the vertical tail area also severed flight control cables. As he began to descend the pilot had his control column fully aft (in his chest) with no effect. Our attitude soon was 25 degrees nose down with airspeed approaching 390 knots. Our rate of descent was unwinding at a very high rate. My control column flopped about in pitch, but limited roll response was there. The pilot gave me "control."

> ***SYSTEMS NOTE.*** *With only the number four hydraulic system pressurized, the right aileron could be moved. However, only one of two actuators is providing the muscle. Two flight spoiler panels, numbers 5 and 6 are also pressurized and responding to control.*

I pushed the throttles forward toward max EPR. This pitched the airplane up. As the airplane attitude reached 25 degrees nose up and the airspeed was slowing below 180 knots I rolled the airplane into a right roll of about 65 degrees which modified the pitch attitude. It became apparent after a few large oscillations that pitch response to power was possible, and that we had to be quick learners. The airplane was seeking to stabilize in pitch where it was trimmed, 270 knots. Calls to air traffic control to request emergency clearance back to Saigon deteriorated rapidly to silence. We were told to descend to 4,000 feet and hold for further instructions. The tower operator and controllers never did understand the severity of our problems; the language barrier was part of the misunderstanding. Our presence was known, but they had no clue of what was taking place.

We struggled at positioning our plane for a 12 mile final. Emergency extension of the landing gear was begun at 4,000 feet and 230 knots. The decreasing airspeed (from 270 knots) and the nose gear extension pitched the airplane down to about 15 degrees below the horizon. At this point I maneuvered the aircraft toward the rice paddies adjacent to the runway. We were at 269 knots on touchdown and stayed on the ground for about 2000 feet; I jumped on the brakes pushing with all the force I could manage, but we had no hydraulics; there was little or no response from that reflexive effort. The airplane hit a stone wall which was about eight feet high, causing us to become airborne to about 150 feet height. Seeing that we were headed for the Saigon River, I pushed the power forward aiding us to reach the opposite bank with about ten feet of margin. All through this, I was thinking, keep the wings level, fly straight ahead and the personnel in the troop compartment might have a chance. The second touchdown was a horrendous crash. The airplane broke up as it slid along the ground. The wings, cockpit area, aft fuselage, vertical tail became individual entities in the final loss of momentum. All electrical power was lost putting the cockpit in darkness; mud and crud covered the windscreens. The cockpit area did a reversal, rolling on its right side, finally stopping with the nose pointing 180 degrees from the direction of intended landing. We believe that it took about 25 to 30 seconds before the cockpit area stopped moving. The pilot's clear vision window was about five feet above ground. Our good fortune was that it did not jam closed. We had five flight crew on the flight deck. The shocked silence of the event of stopping was broken by the copilot in the jump seat who yelled. "Hey, we're stopped; we are going to make it; let's get out of here". My feet were jammed around the rudder pedals; when I popped my seat belt it took more than the usual scrambling to get on my feet. The area immediately aft of the flight engineers position was a solid wall of metal. The flight engineer was jammed in his seat; I wrestled the hardware with him and followed him out the window, the last of the surviving crew. In the debacle of the airplane breakup, mud and dirt covered the cockpit windows. Naturally, without power the last few moments of motion were in darkness. Getting out into strong sunlight contributed to personal disorientation. The first big distraction was the fireball around the wings which had come to rest about 150 yards away. We located the aft fuselage and headed in that direction to give whatever assistance we could. The people in the troop compartment were roughed up, but survived. One of the nurses, Captain Regina Aune, sustained a broken back. She stayed alert until she tried to move her kids, then passed out. The crash site was covered with 40 to 50 helicopters in about ten minutes, a hazardous blessing, to be sure. Of the approximate 390 souls on board, 180 did not survive. Ten of the fifteen flight crew members made it through the ordeal. Two loadmasters, two flight engineers, and one copilot did not. One of the nurses was also a casualty. The surviving flight crew spent the first night in a commercial hotel in Saigon, returning to Clark AFB on a C-141 the next day. After three weeks of interviews by the accident board we were released to return to Travis AFB. We stayed with our families having little contact with anyone until the investigation was completed. Within a year only two of the flight crew remained on flight status.

Q: What is the plus side of flying the C-5; you stuck with it despite this grim experience?

TILL: The C-5 is an excellent air-lifter. It is really a fine machine to fly. And, the C-5B came along soon enough for me to experience a good flying machine get even better. I've had my opportunities to fly several humanitarian missions; hurricane relief for victims of hurricane Hugo, and hurricane Iniki in Hawii. Plus, I had a piece of the airlift action into Mogodischu, Somolia during Christmas week in 1992. So, when one can fly a seriously needed mission with an air-lifter that can do as much as the C-5 it makes sense to stay with it.

Author's Note

After the accident, all C-5 aircraft had a revised locking system for the pressure door installed. Important also is the fact that the USAF investigating team officially praised the efforts of the flight crew for their skill in keeping the aircraft in a survivable attitude for touchdown; and, more importantly, they located the fuselage after escaping from their disconnected cockpit. They then assisted the surviving passengers to rescue vehicles. This prevented total loss of life of those passengers surviving the crash landing. USAF awarded the Air Force Cross and Airman's medals to the pilot and copilot. The remaining crew members received the Airman's medal and other commendations.

Readers may question why it is of value to include this detail of an aircraft accident that would appear to tarnish the "Image and Record" of the C-5A Galaxy; several come to mind.

Consider the following circumstances.

The C-5 airplane had been in service for less than sixty months. Thus, this flight crew's total accumulated flight exposure to the aircraft in training and operational flight precluded extensive familiarity in controlling the aircraft in non-standard configurations. The initial and recurring training cycles required to qualify in the C-5A exposes crew members to degraded normal systems, such as disabling one or two engines, removing functional elements of the electrical or hydraulic systems, and exercising aircraft control following rapid decompression (loss of automatic pressurization), but these system faults are necessarily analyzed separately and result in the conditions where flight control is not severely degraded. Compound failures may occur, but the rarity of that circumstance dictates that training concentrate on satisfactory management of a single system loss. In fact, there is no training accomplished anywhere which presumes loss of all flight controls, thus suggesting to trainees that engine thrust alone will be available for airplane attitude control. (Consider trying to drive your car without a steering wheel.)

Yet for this crew, the absolute worst did happen.

On 5 April 1975 the in-flight failures were immediate, simultaneous, completely original and had been thought impossible. C-5A, 0011, did indeed attempt to self destruct. Loss of all normal flight control in a matter of seconds, combined with rapid decompression, certainly physically and mentally challenged the flight crew. While control of a single aileron and two associated flight spoilers remained, this residual powered control could just as easily have aggravated any attitude control regained through judicious use of symmetric or asymmetric engine thrust. Total appreciation of the severity this emergency presented to this crew cannot be satisfied by merely recognizing the crew for exhibiting crucial skill in recapturing control of this wildly oscillating airplane and retaining the absolute wings level attitude (the only survivable attitude) for a crushing impact with the ground.

Before deciding where to scale this miraculous airplane recovery, consider two similar in-flight emergencies and their results.

In early 1974 a DC-10 operated by Turkish Airways departed Paris for London with 346 persons on board. Shortly after takeoff, apparently a cargo door latch system failed. The loss of the door set in motion a series of conditions which led to loss of pitch control of the aircraft. The aircraft crashed in moments, leading to the total loss of life. Additional details of the accident and the DC-10 early operational experiences can be read in the book, "Destination Disaster", copyrighted and published in early 1976.

A second example of loss of flight control in large airplanes involves Japan Airlines JL 123, a Boeing 747SR on a flight from Tokyo's Haneda airport to Osaka on August 1985 with 520 people on board. At 24,000 feet a failure of the aft pressure bulkhead (very similar to the Saigon crash) occurred, followed by loss of rudder control and the eventual loss of all flight controls. Rolling and pitching cycles occurred which were uncontrollable by the flight crew. Forty six minutes after takeoff the aircraft impacted mountainous terrain with the loss of everyone on board except one infant child. Details of published reports of the accident were published in Flight International dated 14 September 1985.

It is simplistic to suggest these accidents are similar. Certainly, structural failure was a common link, as were the weather conditions. Factors such as background, training, and individual will to overcome totally unforeseen events are considerations that underline broad differences. Yet, on reflection, each of these three events serves to highlight the genuinely positive factors of the Saigon tragedy.

The primary thought that comes to mind is the appreciation for the determined execution of skillful airmanship to recover flight control of an oscillating monster; a truly unrelenting exercise of the will to live. Also, definitely to be appreciated is the remarkable luck of the passengers who not only survived the flight recovery, but also the merciless impact, and the grinding breakup of the fuselage noisily moving to a sledding stop in the quagmire of a Vietnam rice paddy. Survivors, yes, but still in desperate circumstances which would have rapidly become fatal but for the continued heroic efforts of the flight crew and medical crew members.

Dave Wilson

Pilot, C-5 Flight Instructor

Q: Your C-5 flight experience dates to one of the first operational squadrons. Did you request assignment to the C-5 community?

DAVE: Not at all. I really went into that flight assignment kicking and screaming. I was finishing my tour in Vietnam, and I was happily contemplating reassignment to McClellen AFB to fly the WC-135s again. The C-5 was getting off to a rocky start from my perspective. Seeing those airplanes with smoky engines fly logistic missions into Nam, one noted unusual routines like the two minute engine run-up prior to takeoff; flying into the horizon with the gear still extended, and then, Dick Leal's experience at Altus AFB. All this came across negatively to me. Then suddenly, without any "are you interested" questions, came a notice to report to Dover AFB.

Q: What happened at Altus?

DAVE: They lost an engine prior to take-off, literally; the pylon broke and the engine did a small chandelle over the wing.

Q: So you were in the C-5 program with reluctance; what made you change your mind?

DAVE: There is nothing more convincing than performance. I did transition training at Altus, returned to Dover in July 1972, flew the required line flights accumulating time and experience necessary for upgrade to aircraft commander (AC). The first of the three flights that made me realize how truly capable the C-5 is for heavy logistics occurred in October 1972. We airlifted two Chinook helos from Harrisburg, Pa. to Udorn, Thailand. That flight was an eye-opener. Alternate transport for those helos was by train and ship. The second mission occurred the next month in November. President Nixon felt he had a deal with the North Vietnamese. He offered to clear the Hiapong harbor of mines. So, on one day's notice, several C-5s were assigned to airlift CH-53s from NAS Norfolk to Cubi Point in the Philippines via Travis and Wake Island. I happened to be AC on the first C-5. This mission reflected the ability to carry out national policy "ASAP," a mighty strong endorsement for the airplane. The third mission, probably my most satisfying flight sequence ever, was in December 1972. The earthquake in Nicaragua had done substantial damage to that country. I received a call from crew schedules asking if I would go on alert for a mission to Managua in the next few days. Never happen, I thought, as I answered "Sure, why not." MAC would never risk a C-5 into an area where the runways were short and likely damaged. Besides, it was 23 December. The next day I was alerted for immediate departure to Fort Hood, Texas. C-5 with augmented crew (three pilots, two navigators and three loadmasters) picked up hospital equipment, water tankers and medical supplies and flew to Howard AFB; no one lands at night at a damaged airport. At first light on Christmas morning we flew into Managua, Nicaragua. This mission required fifteen sorties and thirty one flight hours. Six sorties air-lifted a total of 844,441 pounds of cargo into Managua. The cargo was two ambulances, one 18 wheeler water truck, five water purification mobile tanks, two D-7 cats, one D-8 cat, three lowboy trailers with tractors, a Jeep, two 19 ton wreckers, one 10,000 kw generator, a half dozen all terrain forklifts, six dump trucks with trailers, and a whole bunch of other "stuff." We carried passengers back to Panama. Our loadmasters gave away every ounce of food we had on board. One passenger, a Miss Joan Atkinson, had been buried in the rubble for several hours after the quake. We flew her back to the States. Our turn around time at Howard, AFB averaged about 2.1 hours; in Managua it was under one hour. This allowed us two sorties per day into Managua. This is one mission that will always be special. One doesn't often get the chance to do your job, serve your country and assist greatly in coping with a human catastrophe at Christmastime. After that mission I became a true believer in the C-5, and was very glad to be here.

Q: Did the field conditions do much damage to the airplane?

DAVE: No. We had to taxi over some rough spots and do some extensive wheel braking. The only maintainance required for the airplane back at Dover was tire changes. After sixteen were replaced, the airplane was back in A-1 status.

Q: Did you have any involvement in "Operation Nickelgrass," the airlift to Tel Aviv?

DAVE: Yes, I flew three missions. For that period the airplane operated with a gross weight restriction of 728,000 pounds. The cargo into Israel was, as expected, ammo, trucks, tanks and other "stuff". We also did a fair amount of backhauling. Both the U.S. and the Israelis need to know what makes enemy war material tick. So we hauled Arab (Russian) tanks, missiles and similar "stuff" to Aberdeen and Wright Patterson for teardown and analysis. The early backhaul missions used a short check list. Load till full, close and go! At times this "stuff" had live ammo and oddball weights. But we delivered.

Q: What's your experience with the in-flight refueling (IFR) capability of the C-5? Do you prefer the KC-135 or the KC-10?

DAVE: No preference at all. IFR became a requirement after "Nickelgrass" experience. The airspeed to refuel from a KC-135 was 252 to 265 knots indicated. It's quite reasonable, the airplane trims nicely. Refueling from the KC-10 the airspeed envelope changed to 275 to 300 KIAS. The 252 KIAS had complications for the KC-10 at its max weights. MAC changed the C-5 nose down trim limit to 2.4 units to accomodate the new flight conditions. We did most refueling at a nominal altitude of 24,000 feet. In 1987 when the operational testing was being completed to gross weights of 840,000 pounds the refueling altitude was lowered to 16,000 feet. These tests did include pilots with very low C-5 experience, newly minted crewmen from Altus AFB. The results were satisfactory. The data is now in the handbook, the 1-1. I do fly with a very slight cross control of 2 to 3 degrees left wing down when refueling behind the KC-10. That center engine exhaust may be contributing to this. Checking with some other pilots I've learned this technique is not uncommon.

Q: What kind of short field take-off and landing experience did you acquire?

DAVE: There are several missions to discuss, but two quick examples. My earliest was at the Reading airport in 1979. The long runway was closed, so we used the 5,000 foot strip. It was raining, and the runway was slightly downhill, yet we stopped without any problem. Routinely, the C-5 can be easily slowed in rollout to taxi off the runway using less than 5,000 feet with only moderate use of brakes. A remarkable example of the C-5 capability occurred at McCall field, a 5,000 foot practice strip near Pope AFB, used for special operations. A C-141 and a C-5 each hauled a load of gear into the field at night. As maneuvers completed, it was raining; time to go back to Pope. The C-141 crew told the operational commander that they could not take their cargo. Field conditions, the wet, short runway, limited their gross weight due to critical field length requirements in case of a rejected takeoff. The C-5 crew volunteered to carry both cargo loads, which they ultimately did. But they had to pull out the performance data and convince the skeptical C-141 crew that it was possible.

Q: Didn't you ever have any really "fun" missions?

DAVE: Oh certainly. In 1976 we had to haul Bell helicopter gun-ships and slicks from Amarillo, Texas to Teheran, Iran non stop. We would meet three tankers off Gander and three more off Rota, Spain making it an eighteen hour flight. The contract fuel operator at Amarillo hated to see our missions come to an end. They had been used to servicing T-38s and T-39s. We roll in and ask for 230,000 pounds of fuel and their adding machines go into overdrive. Needless to say, we got fantastic service. The nice addendum to that cargo was several cases of Cours beer that we were able to enjoy during turn crew rest in the desert. We made friends by the dozen. Then, the opportunity to fly the first C-5 with the prototype wing gave us a taste of what could be called legal abuse testing. In 1981 the airplane with the new wing was at Marietta for fuel tank resealing. With rework completed we were there to bring the airplane to Dover if the flight check proved satisfactory. The fuel load was maximum, 332,000 pounds. At 10,000 feet we did three maximum angle sideslips left and right; we cruised at maximum altitude for five hours to cold soak the fuel, then maneuvered through three pull-ups to 2.25g . Returning to the field we did three hard landings. (Talk about a reversal of form!) Yes, there was evidence of fuel outside the tanks. The airplane stayed at Marietta for a while.

Q: Dave, there is nothing like an airborne emergency to make a lasting impression. You must have had several in 16 years of C-5 flight, any really serious events?

DAVE: That subject has a mixed response. I never had to lower the landing gear by emergency methods for about seven years; on the other hand, early in the program getting all the gear doors to close on initial retraction was a constant an irritant. In my first 700 hours I did have to shut an engine down at least a half dozen times in flight. But my worst engine problem was during takeoff on 8L from Honolulu, just prior to go speed. The number three engine fan bearing froze. N-1 stopped, as well as, the low-pressure turbine. The resulting compressor stall was ferocious; the scanner looking at the engine swore it was shaking violently enough to depart the wing. End result was a seven day layover awaiting repairs. Other system problems were mainly very minor in comparison.

Q: Did the airplane instrumentation cause problems? What about vertical scale tapes?

DAVE: I loved the vertical instruments; my only complaint has to do with the flight director system in general; it is very "unfriendly" and not versitile. It and the expanded scale attitude indicator (ADI) were a constant source of irritation. But pilots always adapt; they are the ultimate practioneers of common sense.

Q: Did you have any experience with special operations, specifically, using the night vision goggles?

DAVE: Yes, I was heavily involved in this area. I flew the initial demonstration flight for the Joint Special Operations Command (JSOC) in 1981 and worked closely with, and flew many missions for them until I retired six years later. An amusing incident that occurred on the initial flight. We landed at North Field, SC (an abandoned airstrip used for NVG work and airdrop) and met the JSOC J-3, an Army Colonel. We shut down the engines while waiting for the sun to set. He remarked about that, seriously concerned that we would not be able to restart our engines. We stated that the thought never occurred to us. We flew that night and showed "our stuff". We dropped him off after the demonstration. He jumped into his King Air and could not start his engines. Well, you can imagine the significant amount of verbal abuse he received. We never let him forget his "moment-in-the-moon."

Chapter References

Chapter 1

1. At the time, scheduled carriers, as represented by the domestic trunklines and the international carriers, flew set schedules over fixed routes and received a federal subsidy. Supplemental or nonscheduled carriers flew based upon temporary needs and lacked subsidies. Another important category was the all-cargo carriers, which had grown out of the supplemental group. They flew fixed routes and could obtain subsidies. In addition, there were a number of carriers that provided feeder service or served regional areas like Hawaii or Alaska.
2. Congressman Flood's colleagues were especially interested in the military's wasteful practices and relationship with the commercial carriers. Ignoring the need for aircraft specifically designed for military use, they directed in the Senate and House Appropriations Committee reports that the Air Force fully utilizes the service of the commercial airlines. For instance, in 1957 Senator Stuart Symington (D-MO) got into law a requirement that civil air carriers handle 40% of the passenger and 20% of the cargo transportation requirements for MATS in fiscal year 1958. See, US Cong,. House, Committee on Appropations, Department of the Air Force Appropriations of 1957, Hearings before a sub-committee, 84th Cong., 2nd Session,. 1956; US Cong., House Committee on Appropriations, Department of Defense Appropriations of 1957, 84th Cong., 2nd Session, House Report 2104, 1956; US Cong., Senate, Committee on Appropriations, Department of Defense Appropriations for 1958, 85th Cong., 1st Session, Senate Report 543, 1957.
3. Department of Defense, The Role of Military Air Transport Service in Peace and War, February 1960, copy in Office of History, Military Airlift Command, Scott Air Force Base, IL.
4. Robert C. Owen, "The National Military Airlift Hearings of 1960: Doctrinal Victory or Successful Turf Battle?" unpublished study in possession of author.
5. US Cong., House Committee on Armed Services, Special Subcommittee on National Military Airlift, Hearings before Special Subcommittee on National Military Airlift, 86th Cong., 2nd Sess., 1960; Frederick C. Thayer, Air Transport Policy and National Security (Chapel Hill; University of North Carolina Press, 1965), pp. 136-139, 196-199; Airforce Pamphlet 19-2-2, Vol III, 1 June 1964, p.26.
6. Charles E. Miller, Airlift Doctrine, (Maxwell Air Force Base, AL: Air University Press, 1987) p. 276.
7. Robert Frank Futrell, Ideas, Concepts Doctrine: A History of Basic Thinking in the United States Air Force, 1961-1984, (Maxwell Air Force Base, AL : Air University Press, 1989), 2:30.
8. US Cong., House Committee on Armed Services, Special Subcommittee on National Military Airlift, Hearings before Special Subcommittee on National Military Airlift, 86th Cong., 2nd Sess., 1960.
9. Walter L. Kraus and Jose M. Matheson, C-141 Starlifter (Scott Air Force Base, IL: Office of MAC History, 1973), pp. 1-2.
10. The problems of creating a separate Air Force have been detailed in Herman S. Wolk, Planning and Organizing the Postwar Air Force, 1943-1947 (Washington, DC: Office of Air Force History, 1984).
11. Initially "outsized" referred to the ability to carry bulky and heavy pieces of equipment like tanks, missiles, and fighter aircraft. Presently, "outsized" is defined as an item exceeding 828" by 117" by 105" high. Oversize is any item that exceeds the usual dimensions of a 463L pallet (104" by 84").
12. MATS Commander to Headquarters USAF, "Qualitative Operational Requirement for Logistic Aircraft Support System," 9 October 1961, Office of MAC History.
13. History of MATS, July 1964-June 1965, pp. 169, 170, 172, Office of MAC History.
14. This had already been percolating for more than at year before it was issued. See History of MATS, July 1962-1963, pp. 247-49.

15. Department of the Air Force, "SOR 214 for Heavy Logistics Weapons System," 25 March 1964; "MATS C-5A Programming Plan," 15 June 1965; History of MATS, July1962-June 1963, pp. 256-258, all in Office of MAC History; Miller, Airlift Doctrine, p. 289.
16. On the aircraft produced by these companies see, Peter M. Bowers, Boeing Aircraft Since 1916 (London, England: Putnam Aeronautical Books, 1968); Renee S. Francillon, McDonnell Douglas Aircraft since 1920, (London, England: Putnam Aeronautical Books, 1979); Renee S. Francillon, Lockheed Aircraft Since 1913 1916 (London, England: Putnam Aeronautical Books, 1982.)
17. "CX-HLS Study Winners," Aviation Week and Space Technology, 8 June 1964, p. 18.
18. History of MATS, July 1964 – June 1965, p. 172, Office of MAC History.
19. Berkley Rice, The C-5 Scandal (Boston, MA: Houghton Mifflin Co., 1971) pp. 5-6.
20. Fortune, February 1965, p. 52
21. Martin Caiden, The Long Arm of America: The Story of the Amazing Hercules Air Assault Transport and Our Revolutionary Global Strike Forces (New York: E.P. Dutton and Co., 1963); Sam McGowan, The C-130 Hercules: Tatical Airlift Missions, 1958-1975 (Blue Summit, PA: Aero Books, 1988).
22. On the C-141, see Harold H, Martin, Starlifter (Brattleboro, VT: Greene, 1972); Kraus and Matheson, C-141 Starlifter.
23. Paul Eddy, Elaine Potter, and Bruce Page, Destination Disaster: From the Tri-Motor to the DC-10, The Risk of Flying (New York: Quadrangle Books, 1976), pp. 104-106.
24. On this approach toward contracting see, Arthur M. Schlesinger, Jr., A Thousand Days: John F. Kennedy in the White House (Boston: Houghton Mifflin, 1965); Alain C. Ethoven and K. Wayne Smith, How Much is Enough? Shaping the Defense Program, 1961-69 (New York: Harper and Row, 1971); Robert S. McNamara, The Essence of Security: Reflections in Office (New York: Harper and Row, 1968); William W. Kaufman, The McNamara Strategy (New York: Harper and Row, 1964).
25. For insightful discussion of how Congress has changed the manner in which it deals with defense issues see, James M. Linsdey, "Congress and Defense Policy: 1961-1986," Armed Forces and Society, 13 (Spring 1987): 371-401; Edward A Kolodziej, The Uncommon Defense and Congress, 1945-1963 (Columbus: The Ohio State University Press, 1966); Craig Liske and Barry Rundquist, The Politics of Weapons Procurement: The Role of Congress (Denver: University of Denver Press, 1974); Raymond H Dawson, "Congressional Innovation and Intervention in Defense Policy: Legislative Authorization of Weapons Systems," American Political Science Review, 55 (March 1962): 42-57.
26. Joint Economic Subcommittee on Economy in Government, Hearings on Economic of Military Procurement, June 1969, Part 1, p. 116.
27. History of MATS, July 1964-June 1965, pp. 171-73.
28. Bernard and Fawn M. Brodie, From Crossbow to H-Bomb (Bloomington: Indiana University Press, 1973 ed.), pp 281-85; Gerard H. Clarfield and William M. Wiecek, Nuclear America: Military and Civilian Nuclear Power in the United States, 1940-1980 (New York: Harper and Row, 1984), pp. 232-237; David Schwartzman, Games of Chicken: Four Decades of US Nuclear Policy (New York: Praeger Publishers, 1988), pp. 67-99.
29. General Howell M. Estes, Jr., "The Revolution in Airlift," Air University Review, 17 (March-April 1966): 2-15 quote from p. 4.
30. Ibid., p. 6.
31. Ibid., p. 6-9.
32. General Howell M. Estes, Jr., "Modern Combat Airlift," Air University Review, 20 (September – October 1969): 12-25.

Chapter 2

1. Lockheed Georgia Proposal Documentation.
2. Engineering Lectures of F.M. Wilson.

Chapter 3

1. Quoted in John Newhouse, The Sporty Game (New York: Alfred A. Knopf, 1982), p. 4.

2. Senate Committee on Armed Services, Hearings, Weapons Systems Acquisition Process (Washington, DC: Government Printing Office, 1972), 92nd Sess., 1st Sess., p. 152.
3. See Harold Asher, Cost-Quality Relationsuhips in the Airframe Industry, Study No. R-291 (Santa Monica, CA: RAND Corp., 1958); Jacques S. Gansler, The Defense Industry (Cambridge, MA: The MIT Press, 1980).
4. Quoted in Berkeley Rice, The C-5A Scandal: An Inside Story of the Military-Industrial Complex (Boston: Houghton Mifflin, 1971), p. 19.
5. Lyndon Baines Johnson, The Vantage Point: Perspectives on the Presidency, 1963-1968 (New York: Holt, Rinehart, and Winston, 1971), p. 20.
6. Norman Moss, Men Who Play God (London, England: Gollancz, 1968), p. 567.
7. Quoted in Benjamin C. Bradlee, Conversations with Kennedy (New York W.W. Norton and Co., 1975), p. 63.
8. This is well discussed in Charles D. Bright, The Jet Makers: The Aerospace Industry from 1945-1972 (Lawrence: The Regents Press of Kansas, 1978), pp. 70-73.
9. Bright, The Jet Makers, p. 70.
10. Richard Austin Smith, Corporations in Crisis (Garden City, NY: Doubleday and Co., 1963).
11. Paul Eddy, Elaine Potter, and Bruce Page, Destination Disaster: From the Tri-Motor to the DC-10: The Risk of Flying (New York: Quadrangle Books, 1976), pp. 106-107.
12. Baltimore News American 9 February 1969.
13. Fortune, August 1969; Get citation for To Engineer is Human.
14. Lockheed-Georgia Company, "The C-5 Contract: Total Package Procurement," July 1969: Lockheed-Georgia Company, "Background on the C-5 Program," 1971, copies of both in possession of authors; Rice, The C-5A Scandal, pp. 24-25.
15. Aviation Week & Space Technology, 4 October 1965, p. 21.
16. Senate Armed Services Committee, Hearings on Military Procurement for Fiscal Year 1970, Part 2, May-June 1969, pp. 1993-2193.
17. Senate Armed Services Committee, Hearings on Military Procurement of Fiscla Year 1970 (Washington, DC: Government Printing Office, 1969), Part 2, May-June 1969, pp. 2009-2010; Lockheed Corporation, "Synopsis of The C-5A Scandal, p. 28.
18. Rice, The C-5A Scandal, p. 10; Lockheed Corp., "Synopsis of The C-5A Scandal," May 1971, p. 10 copy in possession of authors.
19. See Senate Armed Services Committee, Hearings on Military Procurement for Fiscal Year 1970, Part 2, May-June 1969, pp. 2007-2028, for a discussion of this selection process.
20. New York Times, 3 March 1968.
21. Lockheed Corporation, "Synopsis of The C-5A Scandal," p. 10.
22. Quoted in Lockheed-Georgia Company, "C-5 Contract Weekly Project Status Report," 8 October 1965, copy in possession of authors.
23. Lockheed-Georgia Co., "The C-5 Contract: Total Package Procurement," 1969, pp. 1A-1H.
24. For the Record: The Other Side of the C-5 Story, States Before the Committee on Armed Services, United States House of Representatives (Washington, DC: Lockheed-Georgia Co., 1969), pp. 11-22.
25. Quoted in Rice, The C-5A Scandal, pp. 26-27; Fortune, August 1969; Joint Economic Subcommittee on Economy in Government, Hearings on Economic of Military Procurement, Part 1, January 1969, p. 321.
26. Lockheed-Georgia Co., "C-5 Contract Weekly Project Status Reports," 8 October-30 December 1965.
27. D.J. Haughton's US Senate Bank Committee Report, June 1971.
28. C-5 Program Review by F.N. Wittaker for Secretary of Air Force, July 1969.
29. Marietta Journal, August 1971.
30. Lockheed Engineering Report, LG 80 ER0181, 31 October 1980.
31. LG 80ER 0181 31 October 1980, C-5 Structural Integrity Program, Master Plan, Volume II.
32. MAC Operational History, 1982.

Chapter 4

1. Lockheed Engineering Flight Test Reports.

Chapter 5

Interviews from the flight crew members on the first flight.

Chapter 6

1. C-5 Flight Handbook, T.D. 1C-5A-1, dated 5 October 1997.
2. G.E.'s role in TF-39 development. MC Helmsworth 11/19/1981.
3. Technology Improvements listed in GELAC IDC W. Holland to LE Frisbee Oct. 1966.
4. Author's observation notes.
5. USAF Letter, Lt. Col. P. LaCombe, Deputy Director Public Affairs to G. Larson.

Chapter 7

1. Chaim Herzog, The Arab-Israeli Wars (New York: Random House, 1982), p. 230.
2. A discussion of this effort is present in William B. Quandt, Soviet Policy in the October 1973 War, Santa Monica, CA: Rand Corporation Report R-1864-ISA, May 1976), pp. 18-27.
3. On the efforts of the United States to negotiate a truce to the war see Henry Kissinger, Years of Upheaval, (Boston: Little Brown, 1982), pp. 450-544; Richard M. Nixon, RM: The Memoirs of Richard Nixon (New York: Grossett & Dunlap, 1978) pp. 920-44.
4. Nixon, Memoirs, p 22.
5. General Accounting Office, Airlift Operations of the Military Airlift Command During the 1973-Middle East War, (Washington, DC: Government Printing Office, 16 April 1975), p. 6.
6. This situration is ably discussed in Kenneth L. Patchin, Flight to Isreal: Historical Documentary of the Strategic Airlift to Isreal (Scott Air Force Base, IL: Office of MAC History, March 1974) pp. 1-23.
7. This airlift has been described in Charles W. Dickens, "This Isreali Airlift," Airlift Operations Review, (October 1979): 26-28; Chris J. Krisinger, "Operational Nickel Grass: Airlift in Support of National Policy," Airpower Journal, 3 (Spring 1989): 16-28; Patchin, Flight to Isreal.
8. On the History of this organization see Roger D. Launius, The Military Airlift Command: A Short History (Scott Air Force Base, IL: Office of MAC History, 1988).
9. History of Military Air Transport Service, 1962; History of Military Air Transport Service, July 1964-June 1965, pp. 169, 170, 172.
10. Robert Frank Futrell, Ideas, Concepts, Doctrine: A History of Basic Thinking in the United States Air Force (Maxwell Air Force Base, AL: Air University Press, 1988ed.), 2:637 – 645; Robert H. Rea and Graeme M. Taylor, "The C-5A: A Case Study in Weapon System Development, Part A-Concept Formulation," January 1967; History of Military Airlift Command, July 1969 – June 1970, pp. 85-92; Congressional Record, 5 February 1970, p. S1237; press release, Office of Senator William Proxmire, C-5A, 26 January 1971, all in Office of MAC History, Scott Air Force Base, IL.
11. Kissinger, Years of Upheaval, pp. 491-96; George S. Maxwell III, "Israeli Defense Force (IFD) Logistics in the Yom Kippur War," Theses, Air Force Institute of Technology, 1986, p. 53; General Accounting Office, Airlift Operations of the Military Airlift Command During the 1973 Middle East War, p. 8.
12. General Accounting Office, Airlift Operations of the Military Airlift Command During the 1973 Middle Ease War, pp. 7-9.
13. Kissinger, Years of Upheaval, pp. 509-14.
14. Ibid, pp. 514-15.
15. General Accounting Office, Airlift operations of the military Airlift Command During the Middle East War, pp. 7-9; Maxwell, "Israeli Defense Force Logistics," pp. 54-57.
16. Military Airlift Command Oral History Program, Interview No. 3. Major General William E. Overacker, Chief of Staff, Military Airlift Command (Scott Air Force Base, IL: Office of MAC History, April 1990), p. 5.

17. General According Office, Airlift operations of the Military Airlift Command During the 1973 Middle East War, p. 9; Maxwell, "Israeli Defense Force Logistics," pp. 55; Robert Trimpl, "Interview with General Paul K. Carlton," Airlift: The Journal of the Airlift Operations School, 5 (Winter 1984): 17.
18. Charles E. Miller, Airlift Doctrine (Maxwell Air Force Base, AL: Air University Press, 1988), pp. 340-42.
19. MAC Directorate of Information, The Military Airlift Command's Role in the Israel Airlift of 1973 (Scott AFB, IL: MAC, March 1974), pp. 4-6.
20. Later the definition of outsized cargo came to be any item exceeding 828" by 117" by 105" high. Oversize is any item that exceeds the usual dimensions of a 463L pallet (104" by 84").
21. Maxwell, "Israeli Defense Forces (IDF) Logistics," p. 53.
22. GAO, Airlift Operations of the Military Airlift Command, p. 34.
23. Patchin, Flight to Israel, p. 249.
24. Ibid.,p. 253.

Chapter 8

C-5A World and National Records.

Chapter 9

C-5B System Data.

Chapter 10

Comments by senior management and experienced pilots – E.B. Gibson; LL Frisbee, RB Ormsby E.A. Gustafson, G. Gray; TL Harp; D. Wilson.

APPENDIX A

WEIGHT LIMITS APPLIED TO THE C-5

All airplanes have weight limits. The C-5 category of airplanes have a tabular series of weight limits that must be individually considered and documented to achieve the desired take-off gross weight. The sum of all unit limits defines the gross weight for any given flight.

BASIC DEFINITIONS

Maximum Gross Weight: this is the sum of the operating weight empty, the weight of cargo, and the fuel weight.

Operating Weight Empty: the weight of the empty airplane, plus any special equipment added to perform a mission. It includes fuel that cannot be drained, engine oil, oxygen, crew equipment and the weight of the crew members.

The C-5 maximum fuel capacity is 335,000 pounds. Maximum cargo is 291,000 pounds, with flight restrictions. Thus, there is always a trade-off between fuel and cargo weights, since the "bare bones" C-5's will weigh 324,000 pounds and up. Maximum flight gross weight limits are set for 2.5g; 2.25g; and 2.0g. There are maximum fuel weight limits for landing.

Two gross weight limits are specified for landing. These gross weights will be paired with a maximum allowable sink rate at the moment of touchdown. They are either 6 feet per second (FPS) or 9 FPS.

In the twenty five year life of the C-5 six major events altered weight limits. These are as follows:

a)1963...DOD issues the initial specification;
b)1965...The production contract is issued.
c)1970...The 80% flight limits are imposed.
d)1982...New wing is installed on the C-5A.
e)1983...The C-5B contract is issued.
f)1984...C-5B service configuration defined.

	C-5A Specification	C-5A Initial contract	80% Operation
Max T.O. Gross Wt.	769,000 lbs	769,000lbs.	712,000 lbs.s
Max 2.5g Flight wt.	728,000 lbs.	728,000 lbs.	——
Max 2.25g Flight Wt.	769,000 lbs	769,000 lbs.	——
Max Landing Weight	635,850 lbs	635,850 lbs	635,850 lbs
(Sink rate)	(9 FPS)	(9 FPS)	(9 FPS)
Max Landing Fuel Wt.	318,500 lbs.[3]	318,500 lbs.[3]	318,500 lbs.
(Sink rate)	(6FPS)	(6FPS)	(5 FPS)
Operating Weight	323,904 lbs.	323,904 lbs	351,072 lbs.
Maximum Payload	220,000 (2.5g)	190,000 (2.5g)	146,928 (2.0g)
(g's)	265.000 (2.25g)	235,000 (2.25g)	265,000 (2.0g)
	——	265000 (2.0g)	——
AILERON UPRIG	0 degrees	6 to 12 degrees	6 degrees (ALDCS)

Automatic Lift Distribution Control System (ALDCS)

	C-5A Wing Mod configuration	C-5B Contract configuration	C-5B Service configuration
Max T.O. Gross Wt.	837,000 lbs.	837,000 lbs.	837,000 lbs.
Max 2.5g flight Wt.	769,000 lbs.	769,000 lbs.	769,000 lbs.
Max. 2.25g flt. Wt.	840,000 lbs[3,9]	840,000 lbs.	840,000 lbs.
Max. Landing Wt.	920,000 lbs. (6 FPS)	920,000 lbs. (6FPS)	920,000 lbs. (6FPS)
Max Land Fuel Wt.	332,500 lbs. (6FPS)	332,500 lbs. (6FPS)	332,500 lbs. (6FPS)
Operating Wt. Empty	370,000lbs.[4]	370,300 lbs.[8]	374,000 lbs.[10]
Max Payload (g's)	200,000 (2.5g) 245,000 (2.25g) 275,000 (2.0g)	219,700 (2.5g) 264,700 (2.25g) 294,700 (2.0g)	216,000(2.5g) 261,000(2.25) 291,000(2.0g)
Aileron Uprig	0^0 (ALDCS)	0^0 (ALDCS)	0^0 (ALDCS)

Notes

1. 100% ground handling criteria.
2. MAC operations and LAC engineering coordination required.
3. Does not include ECP's Cargo Fire Suppression System, aft troop kit.
4. Minimum radius turns prohibited. 30 knots limit taxi speed.
5. Maximum gross weight for kneeling and unkneeling.
6. Includes AF negotiated changes at C-5B contract go ahead.
7. Minimum radius turns prohibited, 10 knots taxi speed limit.
8. Contract weight for performance evaluation.
9. Weights for in-service operation and structural design evaluation.
10. Contract configuration owe plus allowance for troop seats, refrigerators and secure voice equipment.

Appendix B

Background Material – Personnel Interviewed

E.B. Gibson

As Chief Advanced Design Engineer for the Lockheed Georgia company he led the team that provided the technical data for the winning competitive bid for the C-5A Galaxy. He was appointed C-5A Chief Engineer and Engineering Program Manager when the award was made to Lockheed. He held the position for four years.

Gibby was initially employed at Lockheed California, at age 19, in 1938. During his first fifteen years at the California division he worked in design on the P-38 fighter, the model 14 and model 18 civil transports, both the civil and military versions of the Constellation, the XFV-1 tailsitter, the P2V Navy patrol plane, and the YC-130.

In 1953 he transferred to the Georgia division where he worked in management positions on the C-130 series transports, the Jetstar, the C-141 Starlifter, and the XV-4A Hummingbird prior to his assignment as Engineering Program Manager of the C-5A program. He retired in 1982 after 44 years of employment with Lockheed.

L.E. Frisbee

Lloyd Frisbee had a combined fifteen years engineering experience with Boeing, Seattle in Flight Research and Aerodynamics and with Northrup in Engineering Experimental Flight Test prior to being hired by Lockheed to assemble an Engineering Flight Test development organization for the Lockheed Georgia Division in 1953. Lloyds tenure at GELAC extended through the development of the C-130A, B, E, the C-141 Starlifter, Jetstar, Hummingbird, and the first three years of the C-5A development and flight test programs. Lloyd transferred to the California division as Director of the L-1011 program at the Burbank n Lockheed in 1981.

E.A. (Eddie) Gustafson

Eddie Gustafson graduated from Aero Tech college in 1941, worked for a small aircraft company in California, then transferred to the Bell aircraft company in Marietta as a design engineer on the B-29 airplane. He then served in the US Army for 2 years; Eddie joined Lockheed in 1951.

Eddie was Assistant Project Engineer for the C-5A responsible for fuselage design during the proposal phase and after contract award. His principal assignment during the design and production of the C-5A was to direct activities of the off-shore design group at Phoenix House in Middlesex, England.

R.B. (Bob) Ormsby

Bob was deputy to the C-5A Engineering Program Manager from 1965 to 1967. He graduated from Georgia Tech in 1945, worked for G. L. Martin and the Bureau of Aeronautics prior to joining the Lockheed Georgia company in 1954. Bob was appointed president of the Lockheed Georgia company in 1976. He joined the Lockheed corporate staff in 1979. He retired in 1986.

E.J. (Ed) Schockley

Ed graduated from Georgia Tech in 1950 after serving with the US Navy during WW II as a pilot , and initially worked with Douglas Aircraft as a flight test engineer. He joined Lockheed in 1953 but stayed in Burbank, California working on the prototype YC-130 airplanes. When the program was transferred to Georgia division Ed moved also. After a series of managerial promotions Ed was tasked as Chief Development Engineer at Marietta during the C-5A development flight and ground testing. Ed retired from Lockheed in 1987 after 3 years as president of Lockheed Ontario.

H. Bard Allison

Bard Allison is a 1953 graduate of the University of South Carolina. He obtained a Masters in mechanical engineering from Georgia Tech in 1959. In his career with Lockheed, he began as a propulsion engineer in the C-130 project. In his 37-year span, he served the company as Chief Engineer, Research and Technology; Director of Engineering; and Executive Director of the C-5 and C-141 programs. His final assignment was Division Executive Vice President and General Manager prior to retirement.

Glenn M. Gray

Glenn, a US Navy pilot with both multi engine and helicopter experience, graduated B of AE honors in aeronautical engineering from Georgia Tech in1955. He joined Lockheed Georgia in 1957 as a test pilot. During the C-5A development testing his primary assignment was to fly ship 6002 in a comprehensive series of performance flight tests. His varied flight testing background included flight testing all models of the C-130, the C-141 Starlifter, the Jetstar, and the XV-4A Hummingbird. Glenn was promoted to Chief Pilot in 1981. He took on added responsibility as Director of Flight Operations in 1983. Glenn retired in1989.

Dave Wilson

Dave enlisted in the US Air Force in 1960 as an airman, completed navigator school in June of that year. His assignments included flying as navigator on B-52Hs. He flew the nav slot on the B-52 that set a worlds un-refueled distance record in 1966. Rutan and Yeager exceeded this mark by flying the Voyager around the world in 1986.

Dave completed pilot flight training and was initially assigned to C-135s. In 1970 he flew Carribous in Vietnam. His next flight assignment was to Dover AFB for transition to the C-5A. He logged 4500 hours in the Galaxy prior to his retirement from active duty. After retirement Dave was a C-5 Flight Instructor in the Dover AFB Flight Simulator.

Ric Johnstone

Ric is currently the Director of Flight Operations at the Lockheed Martin Aeronautical production facility in Marietta, Georgia. He joined the US Air Force in 1970, initially flew C-130s, and KC-135s. He was assigned to the second group of pilots checking out in the C-5A in 1973. After nine years active duty he joined the Lockheed Georgia company while retaining military status in the Reserves. Ric became commanding officer of the 326th Airlift Squadron, a C-5A Reserve squadron based at Dover AFB in 1993, a responsibility he held for three years.

Til Harp

Til enlisted in the USAF as an airman in 1964. He qualified for the Air Force Academy and graduated in 1970. After completing pilot training he was assigned to fly C-141s. September of 1973 he transitioned to the C-5A and flew in type for the next five years from Travis AFB. His progressive responsibilities included obtaining a masters degree, a two year assignment in War Plans Division at MAC headquarters; he then served as Commanding Officer of the Altus AFB training squadron; ADO at Travis AFB, then deputy Operations Group Commander. He was Vice Wing Commander at McConnell AFB prior to his retirement in 1995.

Leo J. Sullivan

Leo was the pilot of the C-5A on its first flight on 30 June 1968. He was first employed at Lockheed California where he flew the P2V. He transferred to the Lockheed Georgia division in Marietta in early 1954 as assistant chief pilot to Bud Martin. Leo was appointed Chief Engineering Test pilot of the newly established group in 1956 and directed this organization during the flight testing of the C-130A, B, E and H aircraft, the C-141 Starlifter, the JetStar and XV-4A Hummingbird airplanes. He flew first flights on all the aircraft during his tenure as Chief Pilot. His flight expertise was evident in his numerous contributions during the early development of these remarkable airplanes.

W.E. (Walt) Hensleigh

Walt was a graduate of Mississippi State University, flew in the Army Air Corps during the last years of WW II. He worked for Fairchild and flew the C-119 and C-123 in test duties. He joined the Lockheed Georgia company in 1957 as an engineering test pilot. He flew performance and flying qualities testing on several versions of the C-130, C-141, the JetStar and the C-5A. He was designated Chief Engineering Test Pilot in July 1968. Walt retired as Director of Flight Operations of the Lockheed Georgia Company in1984.

Lt. Col. Joe Schiele, USAF Pilot

Joe had an extensive career in the US Air Force flying various multi engine aircraft current at that time including the C-124 and the C-133 . He was the USAF pilot on the first flight of the C-5A. Joe is an Associate Fellow of the Society of Experimental Test Pilots.

Ehrhard (Mitt) Mittendorf

Mitt was the flight test engineer on first flights of both the C-5A and the C-141. He studied for two years at Texas A&M prior to enlisting in the US Army in 1942. He completed navigators school in nine months, volunteered for B-17 duty in the European theatre, flew from Grafton, Underwood station106, Mankettering with the 384th Bomb group. Fifteen successful missions over Schweinfurt, Bremerhaven and Berlin led to mission sixteen when the aircraft was shot down. Mitt spent 401 days in a POW camp before rescue. He completed his aeronautical engineering at Texas A&M, worked for Convair for five years, and joined Lockheed Georgia in 1954.

Mitt recalls his assignment as FTE on the first C-141 scheduled to fly on 7 December 1963. The scheduling pace outstripped the maintainence and engineering completion of requirements. When Mitt pointed out to Chuck Whetstone, flight test coordinator, that 17 December was the anniversary of the Wright brothers first flight, approval was obtained to fly on the 17th of December. Even the weather cooperated! Leo Sullivan, Vern Peterson, Bob Brennan and Mitt were the crew for the Starlifter's first flight.

J.R. (Jerry) Edwards

Jerry served in the US Navy from 1951 to 1957. His flight experience was on R7V as flight engineer. He joined the Lockheed Georgia division in 1960 acquiring C-130 and C-141 flight experience prior to his assignment to the C-5A program. He was the flight engineer on the first C-5A flight on 30 June 1968. He has extensive flight test experience on all Lockheed Georgia aircraft including the C-130J.

W.W. Harris

Bill Harris worked for the Lockheed Georgia company from 1955 to 1986. He was involved in flight testing the C-130-A,B,E, H and Tanker configurations. Also, he developed flight test programs for the C-141A and C-5A with a particular expertise in the aerial delivery system (ADS) of all three cargo aircraft.

He began his career in the US Navy checking in as a tail-gunner in TBMs, began his pilot training in mid 1945, earning pilots wings in 1947. In 1950 he separated from active duty, and returned to and graduated from Georgia Tech in 1953 with a Bachelor degree in aeronautical engineering.

Loading a super magnet at O'Hare in Chicago, June 1977. *See page 6.*

C-5 with "yellow paddles" on wing tips, a flutter device. Lockheed photo.

MAC
80214

C-5A modified to carry the NASA Hubble telescope and other oversized equipment. Lockheed photo.

M-60 Tank tied down in a C-5 cargo compartment US Army photo.

Nose section transported from Rhein Main, Germany to Lajes AB.

Portable scissor bridge carried intact by a C-5, ready for immediate deployment.

Tank unloading from a C-5. Lockheed photo.

INDEX

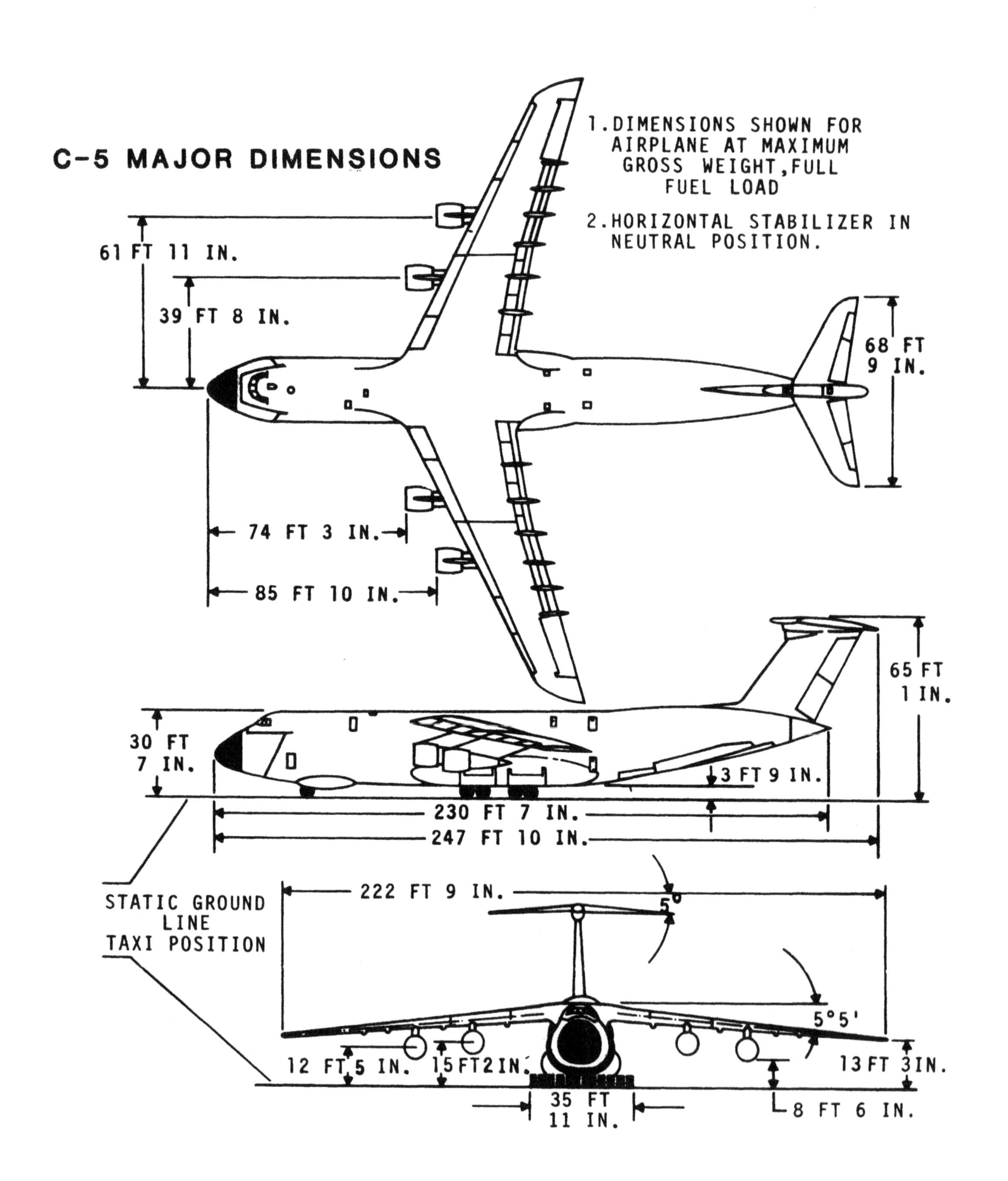
C-5 MAJOR DIMENSIONS
1. DIMENSIONS SHOWN FOR AIRPLANE AT MAXIMUM GROSS WEIGHT, FULL FUEL LOAD
2. HORIZONTAL STABILIZER IN NEUTRAL POSITION.
61 FT 11 IN.
39 FT 8 IN.
68 FT 9 IN.
74 FT 3 IN.
85 FT 10 IN.
65 FT 1 IN.
30 FT 7 IN.
3 FT 9 IN.
230 FT 7 IN.
247 FT 10 IN.
STATIC GROUND LINE TAXI POSITION
222 FT 9 IN.
5°
5°5'
12 FT 5 IN.
15 FT 2 IN.
13 FT 3 IN.
35 FT 11 IN.
8 FT 6 IN.

BASIC DATA	WING	VERTICAL	HORIZONTAL
AREA, SQ FT	620[illegible] *	96[illegible].067	965.824
ASPECT RATIO	7.75 *	1.24	4.74
TAPER RATIO	0.371 *	0.800	0.370
SWEEP AT 25% CHORD	25° ●	34°56′	24°35′
ROOT CHORD, IN.	545.303	371.197	250.160
CHORD PLANFORM BREAK	336.366	——	——
TIP CHORD, IN.	84.036	296.958	92.559
MAC IN. (PROJECTED)	370.523◆	335.452	183.438
AERODYNAMIC SPAN, IN.	2630.437	414.256	811.620
ANHEDRAL	5°5′32″**	——	5°0′
INCIDENCE	3°30′***	——	+4° TO -12°

◆ (TRUE MAC = 371.209) (REF)

● SWEEP OF OUTER WING IN ROOT CHORD PLANE

* IN AERODYNAMIC PLANE

** IN AIRPLANE FRONT VIEW, 3°30′0″ AROUND ROOT CHORD

*** ROOT CHORD

WEIGHT ●	
CONDITION	LB
EMPTY	318,469
DESIGN FLIGHT GROSS (2.5g)	728,000
USEFUL LOAD - DESIGN FLIGHT	409,531
MAXIMUM DESIGN GROSS	769,000
USEFUL LOAD - MAXIMUM DESIGN	450,531

● GUARANTEE FIGURES

(4) TF39 ENGINES

MISCELLANEOUS DATA		AILERON	SPOILERS	TRAILING EDGE FLAPS	LEADING EDGE SLATS	ELEVATOR	RUDDER
AREA SQ FT	INBOARD	——	218.128	493.407	◐ 322.348	180.159	UPPER- 99.
	MIDDLE	——	72.600 △	161.576 □	⊖ 99.147	——	——
	OUTBOARD	252.792	140.004 ▲	336.799	⊖ 227.044	78.515	LOWER-127.
	TOTAL/SIDE	126.396	215.366 ⊘	495.891	324.269	129.337	TOTAL-226
DEFLECTION		UP 25° DOWN 15°	GRD 60° FLT 22.5°	TAKE-OFF 25° LANDING 40°	SEALED 21.5° SLOTTED 22°	UP 25° DOWN 15°	±35°

△ ACROSS WING BREAK (LATERAL CONTROL)

▲ OUTBOARD WING (LATERAL CONTROL)

⊘ ALL SPOILERS (GROUND OPERATION)

◐ INBOARD WING (SEALED)

⊖ WING BREAK TO OUTBOARD ENG

⊖ OUTBOARD ENGINE TO WING TI

□ ACROSS WING BREAK

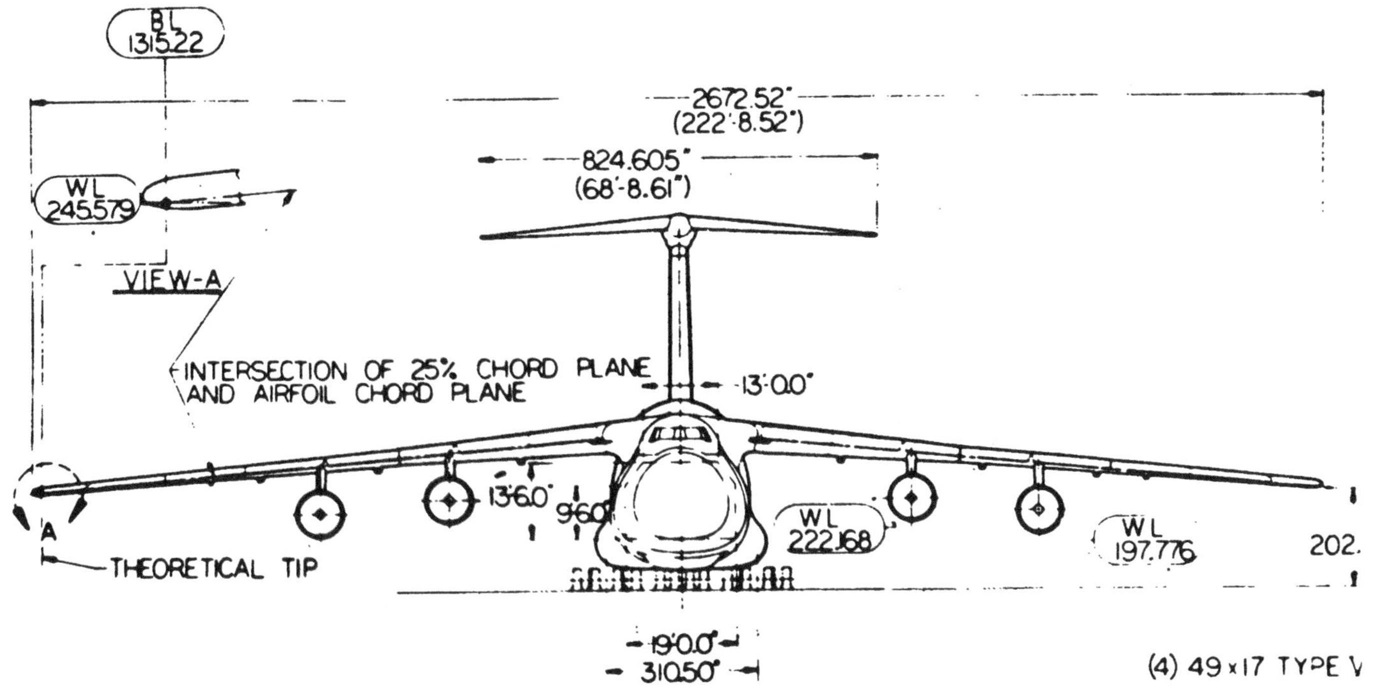

BL 155.25

Printed in the USA
CPSIA information can be obtained
at www.ICGtesting.com
JSHW060045150824
68134JS00031B/2645